A SOCIOLOGICAL LOOK AT BIOFUELS: UNDERSTANDING THE PAST/PROSPECTS FOR THE FUTURE

ENERGY SCIENCE, ENGINEERING AND TECHNOLOGY SERIES

Oil Shale Developments
Ike S. Bussell (Editor)
2009. ISBN: 978-1-60741-475-9

Power Systems Applications of Graph Theory
Jizhong Zhu
2009. ISBN: 978-1-60741-364-6

Bioethanol: Production, Benefits and Economics
Jason B. Erbaum (Editor)
2009. ISBN: 978-1-60741-697-5

Bioethanol: Production, Benefits and Economics
Jason B. Erbaum (Editor)
2009. ISBN: 978-1-61668-000-8 (Online Book)

Introduction to Power Generation Technologies
Andreas Poullikkas
2009. ISBN: 978-1-60876-472-3

Handbook of Exergy, Hydrogen Energy and Hydropower Research
Gaston Pélissier and Arthur Calvet (Editors)
2009. ISBN: 978-1-60741-715-6

Energy Costs, International Developments and New Directions
Leszek Kowalczyk and Jakub Piotrowski (Editors)
2009. ISBN: 978-1-60741-700-2

**Radial-Bias-Combustion and Central-Fuel-Rich
Swirl Pulverized Coal Burners for Wall-Fired Boilers**
Zhengqi Li
2009. ISBN: 978-1-60876-455-6

**Syngas Generation from Hydrocarbons
and Oxygenates with Structured Catalysts**
*Vladislav Sadykov, L. Bobrova, S. Pavlova, V. Simagina,
L. Makarshin, V. Julian, R. H. Ross, and Claude Mirodatos*
2009 ISBN: 978-1-60876-323-8 (Softcover)

**Corn Straw and Biomass Blends: Combustion
Characteristics and NO Formation**
Zhengqi Li
2009. ISBN: 978-1-60876-578-2

**Computational Techniques: The Multiphase CFD Approach
to Fluidization and Green Energy Technologies (includes CD-ROM)**
Dimitri Gidaspow and Veeraya Jiradilok
2009. ISBN: 978-1-60876-024-4

Cool Power: Natural Ventilation Systems in Historic Buildings
Carla Balocco and Giuseppe Grazzini
2010. ISBN: 978-1-60876-129-6

**A Sociological Look at Biofuels: Understanding the Past/Prospects for
the Future**
Michael S. Carolan
2010: ISBN: 978-1-60876-708-3

ENERGY SCIENCE, ENGINEERING AND TECHNOLOGY SERIES

A SOCIOLOGICAL LOOK AT BIOFUELS: UNDERSTANDING THE PAST/PROSPECTS FOR THE FUTURE

MICHAEL S. CAROLAN

Nova Science Publishers, Inc.
New York

NOTICE TO THE READER

The Publisher has taken reasonable care in the preparation of this book, but makes no expressed or implied warranty of any kind and assumes no responsibility for any errors or omissions. No liability is assumed for incidental or consequential damages in connection with or arising out of information contained in this book. The Publisher shall not be liable for any special, consequential, or exemplary damages resulting, in whole or in part, from the readers' use of, or reliance upon, this material.

Independent verification should be sought for any data, advice or recommendations contained in this book. In addition, no responsibility is assumed by the publisher for any injury and/or damage to persons or property arising from any methods, products, instructions, ideas or otherwise contained in this publication.

This publication is designed to provide accurate and authoritative information with regard to the subject matter covered herein. It is sold with the clear understanding that the Publisher is not engaged in rendering legal or any other professional services. If legal or any other expert assistance is required, the services of a competent person should be sought. FROM A DECLARATION OF PARTICIPANTS JOINTLY ADOPTED BY A COMMITTEE OF THE AMERICAN BAR ASSOCIATION AND A COMMITTEE OF PUBLISHERS.

LIBRARY OF CONGRESS CATALOGING-IN-PUBLICATION DATA
Carolan, Michael S.
 A sociological look at biofuels : understanding the past/prospects for the future / Michael S. Carolan.
   p. cm.
 Includes bibliographical references and index.
 ISBN 978-1-60876-708-3 (hardcover : alk. paper)
 1. Biomass energy--Social aspects. 2. Biomass energy--History. I. Title.
 TP339.C376 2009
 662'.88--dc22
                         2009044315

Published by Nova Science Publishers, Inc. ✛ *New York*

CONTENTS

Preface ix

Acknowledgments xi

Chapter 1 Introduction 1

Chapter 2 Better Things for Better
Living...Through Chemurgy 7

Chapter 3 Gasoline Gains Momentum 29

Chapter 4 Ethanol's Comeback 45

Chapter 5 From Food to Feed 61

Epilogue 77

About the Author 81

Index 83

PREFACE

This book concentrates on biofuels and specifically on ethanol within the United States (US), though in the concluding chapter the author expands the discussion to include other agro-based fuels. The focus reflects current production realties. According to the Renewable Fuels Association, some 168 ethanol distilleries in the U.S. produced more than 9.2 million gallons of ethanol in 2008 (up from 6.5 million in 2007). Biodiesel production in the U.S., by comparison, was approximately 492 million gallons in 2007. In a global context, over 16 billion gallons of ethanol were produced worldwide in 2008 compared to approximately 1 billions gallons of biodiesel. Because of this, "biofuel" overwhelmingly means, and will continue to mean for some time, "ethanol" within not only the U.S. but throughout much of the world.

ACKNOWLEDGMENTS

This book represents a multi-year journey. Stumbling upon an article one winter afternoon a couple years back I read in amazement that ethanol was once the fuel of choice among automotive engineers. But the author—Bill Kovarik—was not talking about the 1970s or 1980s. I learned that ethanol—then called ethyl alcohol—seriously challenged gasoline in the early 1900s as *the* fuel of the internal combustion engine revolution. I was hooked. At the time, at least from what I could find, the only scholarly research on the subject was historical. I found it fascinating but I wanted to know more. What I read was descriptive, telling me what happened. But I wanted to know *why* the story of ethanol unfolded as it did. With that I set to work, utilizing my knowledge of sociological concepts to provide the following analysis.

The argument that follows has been "vetting" to various degrees. I have written four peer-review scholarly articles on the subject of biofuels. The ideas and arguments contained within those articles have been refined and revisited in the following chapters. I would like to make note of these other pieces.

Carolan, Michael S. in press, Ethanol's Most Recent Breakthrough in the United States : A Case of Socio-Technical Transition, *Technology in Society*

Carolan, Michael S. 2009. The Cost and Benefits of Biofuels: A Review of Recent Peer-Reviewed Research and a Sociological Look Ahead, *Environmental Practice* 11(1): 17-24.

Carolan, Michael S. 2009 Ethanol versus Gasoline: The Contestation and Closure of a Socio-Technical System in the United States, *Social Studies of Science* 39(3): 421-48.

Carolan, Michael S. 2009 A Sociological Look at Biofuels: Ethanol in the Early Decades of the 20th Century and Lessons for Today, *Rural Sociology* 74(1): 86-112.

I would also like to take this opportunity to acknowledge a couple people who helped make this book possible. My parents have been a tremendous supporting force throughout my life. You've given me the confidence to do what I do today—thanks. Thanks also to Nova Science Publishers for taking a chance on a young scholar. And then there's Nora: for putting up with me as I work through a problem; for not getting mad when I wake up at 4 in the morning to write (the sound of the coffee machine must wake you); and for taking the time to read through previous drafts. Thank you.

In: A Sociological Look at Biofuels
M. S. Carolan, pp. 1-6

ISBN: 978-1-60876-708-3

Chapter 1

INTRODUCTION

We've all read the headlines. We know our thirst for oil is growing faster—four times faster in fact—than the volume of new oil reserves. If we don't do something, and do it quickly, we'll soon run out; a point recently sharpened by the reality of global climate change, growing calls for energy independence, and instabilities of the price of gasoline.

One often-discussed solution to these problems has been the use of biofuels; though the reviews on these "green" fuels have been, to put it mildly, mixed in recent years. Biofuel is an umbrella term that encompasses a variety of agro-based fuels, from biodiesel (commonly made from either palm or vegetable oil) to corn ethanol, sugarcane ethanol, cassava ethanol, and cellulosic ethanol. This book concentrates specifically on ethanol within the United States (US), though in the concluding chapter I expand the discussion to include other agro-based fuels. The focus reflects current production realties. According to the Renewable Fuels Association, some 168 ethanol distilleries in the U.S. produced more than 9.2 million gallons of ethanol in 2008 (up from 6.5 million in 2007).[1] Biodiesel production in the U.S., by comparison, was approximately 492 million gallons in 2007.[2] In a global context, over 16 billion gallons of ethanol were produced worldwide in 2008 compared to approximately 1 billions gallons of biodiesel. Because of this, "biofuel" overwhelmingly means, and will continue to mean for some time, "ethanol" within not only the U.S. but throughout much of the world.

Yet ethanol also has a past that, while often forgotten, is instructive in giving us a yardstick against which we can compare recent trends. Growing up in Iowa,

---

[1] http://renewablefuelsassociation.cmaill.com/T/ViewEmail/y/698C04744910BF06

[2] http://www.afdc.energy.gov/afdc/fuels/biodiesel_production.html

the largest corn producing state in the U.S., I never knew a time when ethanol was not available as a fuel additive at the gas station. For many people in the U.S., however, their experience with ethanol blends (E10 and E85 being the most common) goes back only a couple years. Even the Lieutenant Governor of Iowa, Patty Judge, during the Governor's Inaugural Address in January, 2007, showed a degree of naivety about ethanol's past, remarking: "Biobased industry is just beginning. Who would have dreamed even a few years ago that we could power our cars and trucks with corn and soybeans [...]?" Most would be surprised to learn what has been called today's "ethanol juggernaut"[3] follows an earlier period when ethanol—known then as ethyl alcohol—looked to surpass gasoline as the fuel that would power the then-fledgling automobile.

As a social scientist, I am interested in providing more than a descriptive historical account of biofuels. My goal is not only getting the facts right but also highlighting patterns and irregularities, which will allow me to make some educated projections about ethanol's future. Overlaying the success of ethanol today against its experience from nearly a century ago is a useful way to enrich our understanding of both eras. The story that follows is one of David versus Goliath and, later, Goliath versus Goliath, concluding with a cautionary word about the future.

There are significant pressures pushing back against ethanol. Ethanol has been harshly criticized for, among other things, driving up food prices, its negative effects on biodiversity, and its marginal net energy gains over gasoline. Yet production capacity of this fuel continues to increase. To understand why this is requires an understanding of certain structural and technological realities, which give ethanol "momentum". To give just one example: the Iowa Political Imperative. Any presidential candidate who wants to win the White House knows they must have a good placing in the Iowa caucuses—the first show in the presidential primary season. To reject ethanol is political suicide in this state. Barack Obama understood this, which helped him win Iowa (and later gave him the coveted endorsement of the American Corn Growers Association). John McCain, in contrast, a long-time ethanol critic, tied for third in Iowa (his change of heart towards the fuel six months before the caucus, when he suddenly became an ethanol advocate, turned out to be too-little-too-late for ethanol proponents). And now Tom Vilsack, the former governor of this state, holds the title of Secretary of Agriculture. Structural realities like this are detailed in the following chapters, and will continue to make "ethanol" synonymous with "biofuel" for years to come.

---

[3] http://www.usatoday.com/news/washington/2007-05-01-3741370429_x.htm

Chapter 2 begins the story, which I start in the mid-1800s with the introduction of a "sin" tax on alcohol. The time during which this tax remained in force—from 1862 to 1906—marked a period of considerable expansion for the oil industry. By the time the tax was repealed on industrial forms of alcohol in 1906 the oil industry was already well-entrenched. Yet, even with this head-start going into the twentieth century, ethyl alcohol (ethanol) remained popular among scientists and automotive engineers. As argued in 1925 by chemist M.C. Whitaker to the Chemists Club of New York:

> "Composite fuels made simply by blending anhydrous alcohol with gasoline have been given most comprehensive service tests extending over a period of eight years. Hundreds of thousands of miles have been covered in standard motor car, tractor, motor boat and aeroplane engines with highly satisfactory results. [...] Alcohol blends easily excel gasoline on every point important to the motorist. *The superiority of alcohol fuels is now safely established by actual experience*" (my emphasis) (Whitaker 1925, quoted in Hixon 1933: 1).

The most significant organizational push behind ethyl alcohol came from what was called the chemurgist movement. Reaching its height in the 1930s the chemurgist movement—which included among its initial supporters Henry Wallace, Henry Ford, and Thomas Edison—signified a high-water mark for ethanol fuel blends. Chapter 2 describes these early glory days for ethyl alcohol, before gasoline became "locked in", structurally speaking, as *the* fuel of the twentieth century.

Chapter 3 describes the downfall of ethyl alcohol. Internal memos from 1921 indicate that General Motors (GM) began planning for a total switch from oil based fuels to alcohol fuels (Kovarik 1998). That same year, the top engineer for GM, Thomas Midgley, drove an automobile from Dayton, OH to the annual Society of Automotive Engineers meetings in Indianapolis using a 30 percent alcohol fuel blend. At the meeting Midgley proclaimed: "Alcohol has tremendous advantages and minor disadvantages, [the former include] clean burning and freedom from any carbon deposit [...] [and] tremendously high compression under which alcohol will operate without knocking" (as quoted in Kovarik 1998:14). Later that year the same engineer discovered lead's anti-knock properties and promptly forgot about ethanol as an inexpensive fuel additive. What caused ethyl alcohol to so-mightily-fall in the decades leading up to World War II? No one factor explains gasoline's victory over alcohol. This chapter gives the reader an understanding of the causal complexity that led to gasoline's near-total market domination by 1940.

Chapter 4 picks up the story with ethanol slowly regaining its organizational feet in the 1970s. The fuel reemerged in the public consciousness during the OPEC oil embargo. Since then it has had a slow ascendency to its heights of today. Chapter 4 discusses what scholars of technological change would call ethanol's "incubation space"—a sphere protected from market forces that allows for institutional learning as well as the formation of network and scale economies. The last decades of the twentieth century marked a period during which ethanol received tremendous support from non-market forces. For example, the General Accounting Office of the U.S. calculated that while total ethanol subsidies in the late 1990s were less than those provided to fossil fuels and nuclear power they exceeded all government subsidies when calculated in per-unit energy terms (GAO 2000). This chapter discusses the formation of these spaces of "incubation" and the larger network effects that government subsidies had on the ethanol industry.

This chapter also examines the effects of changing geo-political realities in a post-9/11 world on ethanol production and consumption. Calls for energy independence and the development of renewable fuels, unstable fuel prices, and heightened concerns over global climate change have shown a spotlight on ethanol. Baylor University, for example, was recently awarded a $492,000 grant by the U.S. Department of Agriculture (USDA) to assist in their research on cellulosic ethanol (Biofuels Digest 2008). Other recent USDA grants into the study of biofuels include an $840,000 award to Washington State University for research into phenols in poplar trees (phenols have similar properties to petrochemicals) and a $50 million award to Michigan State University to further their research into ethanol (Biofuels Digest 2008). In 2008, Iowa State University received a $944,000 grant from the US Department of Energy (DOE) to support a project that uses pyrolysis, gasification and nanotechnology based catalyzation to produce ethanol (Biofuels Digest 2008). Corporations have likewise began to fund research into biofuels: British Petroleum, $500 million to a UC-Berkeley lead consortium; Exxon Mobile, $100 million to Stanford University; Chevron, $25 million to UC-Davis; Conoco Phillips, $22.5 million to Iowa State University; Chevron, $12 million to Georgia Institute of Technology; Chevron, $(amount not disclosed) to Texas A & M University (Sheridan 2007). Even Wal-Mart is funding ethanol research, donating $369,000 to the Arkansas Biosciences Institute at Arkansas State University to complement a $1.48 million U.S. Department of Energy grant to support cellulosic ethanol production research at the university (Christiansen 2008).

There is a term in economic and social theory that is relevant to this analysis: path dependence. Explaining path dependence the social historian Mahoney (2000: 508) writes:

"Self reinforcing sequences often exhibit what economists call 'increasing returns.' With increasing returns, an institutional pattern—once adopted—delivers increasing benefits with its continued adoption, and thus over time it becomes more and more difficult to transform the pattern or select previous available options, even if those alternative options would have been more 'efficient.'"

The logic of path dependence enters into this story after one understands how structural realities—such as production and distribution infrastructures, federal subsidies and expert systems that are designed around this fuel—create a type of socio-material momentum. Chapter 4 discusses the current ethanol juggernaut in terms of how it is being "pushed" forward by this momentum.

I have been referring to ethanol up until this point in rather monolithic terms. In reality, however, not all ethanol is equal. What it is made of, how it is made, and where it is made all affect the social and environmental impacts associated with this fuel. Chapter 5 discusses some of these complexities.

While giving public lectures on ethanol I have seen firsthand the passions that this subject evokes, among both its proponents and antagonists. I have heard critics of this fuel liken corn producers to a type of welfare recipient—a not-too-veiled reference to the subsidies that have long propped up the ethanol industry. And I have heard corn growers describe ethanol critics as ideologues whose beliefs have no grounding in science. To temper these emotions I have found it useful to enter into the discussion of ethanol's costs and benefits by looking at what the science says on the subject. I therefore begin Chapter 5 with a review of recent research into the virtues and drawbacks of various biofuel forms. The findings of this review are mixed. They show that some biofuels—namely, those produced from feed-stock versus food-stock—are clearly better than others. In short: 2nd generation biofuels appear to avoid many of the costs that are associated with 1st generation fuels. This is where the sociological grounding provided in earlier chapters becomes important. While more novel biofuels may come with fewer costs, certain sociological and economic factors may keep us from making a quick transition over to these 2nd generation fuels. Chapter 5 concludes with some informed speculation about the future of biofuels.

Before concluding I offer the following disclaimer: neither ethanol proponents nor critics will be entirely satisfied with what I have to say on the subject. While I am highly critical of this fuel I also see ethanol—particularly cellulosic ethanol—as a potentially valuable substitute to wean us from our addiction to "the black gold". But I am not a cheerleader for this fuel either.

Consequently, those who believe ethanol—and biofuels more generally—could ever fully replace oil will not like how this story ends.

REFERENCES

Biofuels Digest. 2008. "USDA awards $492,000 to Baylor to study cellulosic ethanol inhibitors", May 23, < http://www.biofuelsdigest.com/blog2/2008/05/23/usda-awards-492000-to-baylor-to-study-cellulosic-ethanol-inhibitors/> last accessed September 22, 2008.

Christiansen, R. 2008. "Wal-Mart donates $369,000 for ethanol research", *Ethanol Producer* October, <http://www.ethanolproducer.com/article-print.jsp?article_id=4777> last accessed September 22, 2008.

General Accounting Office (GAO). 2000. *Tax Incentives for Petroleum and Ethanol Fuels.* GAO/RCED-00-301R. http://www.gao.gov/new.items/rc00301r.pdf, last accessed June 29, 2008.

Hixon, R.M. 1933 "Use of Alcohol in Motor Fuels: Progress Report No. 6," Iowa State College, Ames, Iowa, May 1.

Kovarik, Bill. 1998. "Henry Ford, Charles F. Kettering and the Fuel of the Future," *Automotive History Review* 32: 7-27.

Kovarik, Bill. 2003. "Ethyl: The 1920s Environmental Conflict over Leaded Gasoline and Alternative Fuels," Paper to the American Society for Environmental History, Annual Conference, March 26-30, Providence, RI, http://www.runet.edu/~wkovarik/papers/ethylconflict.html, last accessed December 7, 2007.

Mahoney, J. 2000. "Path Dependence in Historical Sociology," *Theory and Society* 29:507–48.

Sheridan, C. 2007. "Big Oil's Biomass Play," *Nature Biotechnology* 25(11): 1201-3.

Whitaker, M.C 1925. "Alcohol for Power," Lecture given to Chemists Club, New York, Sept. 30.

In: A Sociological Look at Biofuels
M. S. Carolan, pp. 7-28

ISBN: 978-1-60876-708-3
© 2010 Nova Science Publishers, Inc.

Chapter 2

BETTER THINGS FOR BETTER LIVING…THROUGH CHEMURGY

It's hard not to look upon history and interpret it through the eyes of the present. This is perhaps no more true than with oil. Given how it has seeped its way into almost every facet of our lives it is hard not to see its ascendency as inevitable. Yet oil's value took time to accrue. As described by Sidney Bolles in 1878 (p. 772): "Until between 1850 and 1860 the finding of oil in this country [U.S.] was scarcely ever viewed otherwise than with indifference or annoyance. Its appearance in salt-springs of Ohio and elsewhere proved very detrimental to the interests of the salt-boilers, and on that account the sight and smell of it were detested." Conversely, ethanol—known then as ethyl alcohol—had wide popularity. It was clean burning and thus produced no noxious fumes or smoke upon combustion. It was easily extinguished with water. And, arguably its most valuable characteristic, it could be made at home with renewable resources.

In 1860, 13,157,894 gallons of alcohol were burned in the US for lighting (Herrick, 1907). The Revenue Act of 1862, however, would alter alcohol fuel's trajectory for decades to come. The Act was passed to help generate revenue to pay for the Civil War. In addition to creating the Internal Revenue Service, it also created a "sin tax" on all alcohol. The alcohol tax began at 20 cents per gallon in 1862 and rose to $2.08 per gallon by 1864 (Herrick, 1907). While there was a temporary alcohol tax imposed between 1814 and 1817 to help fund the War of 1812, and an earlier short-lived "Whiskey Tax," alcohol had largely been untaxed before the Revenue Act of 1862 (Nelson, 1995).

The tax brought about by the Revenue Act of 1862, however, made no distinction between alcohol for drink and what is known as "denatured" or "industrial" alcohol (alcohol made unfit for consumption). (Wood alcohol, known as methyl alcohol, was not taxed because it cannot be ingested nor is it flammable therefore excluding it as a fuel.) While there was some debate about excluding industrial alcohol Congress ultimately voted to tax alcohol in all its forms. The science of denaturing was still not well understood, leading some to question whether it represented an actual chemical fact. As one chemist explained a century ago: "[A]t the time denaturing was not an established fact as it is now" (Herrick 1907: 7). Before the tax, alcohol sold for approximately 50 cents a gallon, which was about half the price of lard oil and a third the price of whale oil (both major fuel sources at the time) (Berton et al. 1982). After 1864, alcohol rose to over $2.50 a gallon. For a point of comparison: the average daily wage in the manufacturing industries in 1860 varied between $.90 (cotton manufactures) and $1.78 (stove foundries) (Long 1960: 69). The tax therefore effectively killed demand for this fuel, save for what was produced utilizing home stills hidden from the watchful eye of the federal government.

While industrial alcohol continued to be taxed in the U.S. other countries began passing legislation to encourage its use. Germany, for example, passed a "denaturing law" in 1887, which exempted industrial alcohol from taxation. Similar legislation was passed in England, the Netherlands, and France—indeed, throughout much of Europe—before the turn of the century (Long, 1906). This legislation appears to have been successful. For example, by the turn of the century between five and six percent of all German potatoes were going into the production of alcohol (*Science News Letter*, 1928). In 1901, the total production of alcohol in Germany pushed past112 million gallons (Baskerville, 1906). By 1903, the use of alcohol to power internal combustion engines in Germany exceeded the one million gallon mark, with the government calling for production of alcohol fuel to move past three million gallons (*Science News Letter*, 1928). Similarly, while Britain possessed tremendous coal reserves (by World War I Britain produced approximately 25 percent of the world's coal [Jones, 1978]), it could not claim any domestic oil fields of significance. Thus, while the U.S. was burning kerosene in the late 1800s much of Europe and England were utilizing alcohol.

In 1902, a major exhibit was held in Paris where alcohol fuel was celebrated. Spectators were treated to demonstrations of machines powered by this renewable fuel, from automobiles, to stoves, lamps, farm machinery, and boats (Kovarik, 1998). In Germany, the military was employing a fuel blend containing 80 percent alcohol and 20 percent benzol (Hamlin, 1915). Similarly, by 1906 approximately 10 percent of all engines built by Deutz Gas Engine Works of Germany were designed to run on denatured alcohol (Kovarik, 1998). In Greece, the use of alcohol fuel had become so great that the government was forced to impose an alcohol fuel tax to compensate for the loss of revenues from petroleum taxes (Berton et al. 1982). At the same time, calls were being made by popular automotive magazines to the Royal Automotive Club of London to provide a £50,000 award to encourage research into the alcohol-powered internal combustion engine (Popular Mechanics 1912: 565). Creating a fuel "dependent upon agriculture", it was argued, would reduce "the extremely serious risks to all road locomotion arising during the event of war" (p. 565).

Eventually, the alcohol industry in U.S. received the break it needed. In 1906, a bill to repeal the tax on industrial alcohol was passed. The bill had wide public support (although it was noted in the Senate that "the Rockefellers" were firmly against the tax repeal [Kovarik, 1998]). President Theodore Roosevelt (the famous Standard Oil foe) supported the bill as did the Temperance Party (hoping to find some virtue in an otherwise evil product), the president of the Automobile Club of America, and most auto manufacturers (Kovarik, 2003). Following the repeal the price of industrial alcohol dropped to roughly 30 to 35 cents a gallon, where it remained in the ensuing years (McCarthy, 2001). Yet this was still higher than the price of gasoline, which cost approximately 15 cents a gallon (Baskerville, 1906).

There were a variety of reasons behind why alcohol cost more than gasoline. Contrary to the standard microeconomic myth, price is an effect of much more than supply and demand. For example, fuel alcohol in Cuba was selling for roughly 10 cents a gallon (Herrick, 1907). In Germany, the price per gallon of alcohol was 13 cents (McCarthy, 2001). One production imposed cost in the U.S. was the denaturing requirement, which involves making the fuel undrinkable and thus exempt from the aforementioned "sin" tax. While no firm numbers exist on the subject, it is fair to say that the cost of alcohol fuel would have been considerably less had the denaturing requirement not existed. At the turn of the century, denaturing was accomplished by mixing 90 percent ethyl alcohol—that is, alcohol made from grain, potatoes, beets, corn, and the like—with 10 percent of methyl alcohol—namely, wood alcohol (Wright, 1907). While the price of industrial alcohol at the time hovered around 35 cent, methyl alcohol cost well over a dollar a gallon (Baskerville, 1906). As one chemist pointed out at the time: "It is somewhat unfortunate that wood alcohol remains still the chief material used in the denaturing process, since when used in so large a proportion as 10 per cent, it is a considerable expense …" (Gradenwitz, 1907: 393). (It is worth noting that to support its potato industry Czechoslovakia in the 1930s mandated by law that all of its automobiles run on a 20 percent alcohol blend using non-denatured—which is to say drinkable—alcohol [*Science News Letter* 1933].) Cultural and political realities within the U.S., however, during an era when the Temperance

Movement was its most influential, would never have allowed booze to be sold at gas pumps.

Then there are the efficiencies created by the oil distribution network, which was assembled during the half century when industrial alcohol was heavily taxed. Before the automobile reached the masses, refined oil—in the form of kerosene— was initially used for lightening. Rockefeller's early attention was thus directed at building a kerosene empire. During the latter decades of the 1800s he purchased bulk kerosene supply stations and wholesaling facilities in major cities. He also shipped kerosene by railroad, benefiting from earlier-negotiated low rail shipping rates. This, in addition to legally questionable business practices (such as discriminatory pricing policies and suppression of competition through bribes and mergers), created certain "efficiencies" for gasoline that Rockefeller himself could not have predicted. With the mass production of the automobile, a massive distribution system, which could easily be converted to fit the needs of gasoline, was already in place. The price of gasoline at the pump would have been considerably higher had this network not already been established when Ford began rolling out en mass the Model T in 1908.

In contrast, while Rockefeller was building a network that would eventually bring gasoline to many of the country's gas stations, there was no comparable infrastructural buildup for alcohol fuel. At the dawn of the twentieth century, the only distribution system in place for agricultural commodities was designed for the purpose of selling grain as food for humans and feed for animals. This fifty year head-start gave gasoline a distinct market advantage as alcohol proponents found themselves starting from square one once the tax on industrial alcohol was lifted in 1906.

The oil industry in the U.S. also had a bit of ecological luck on its side. The first commercial well was brought online in 1859 by E. L. Drake in Pennsylvania. Soon oil wells throughout Western Pennsylvania were established (just at the time that the 1862 alcohol tax was brought into effect). Between 1860 and 1862, oil production in Pennsylvania increased from 2,000 barrels to 3 million (Hordeski, 2007). The "ecological luck" I speak of lies in what these early fields yielded upon refining: a very high percentage of lighter petroleum products—namely, kerosene. In the early 1870s, refiners were recovering about 75 percent kerosene from the raw crude (Jones, 1978). This is important because at the time there was only a market for this "lighter" crude. Fuel oil and diesel, which are derived from the heavier elements of crude, were not yet in wide use. As a result, these early oil fields were immensely profitable, or at least more profitable than they would have otherwise been had they yielded a higher percentage of heavier petroleum products.

Although the company was not formally established until 1870, the origins of Standard Oil can be traced back to 1865, when Rockefeller began to buy up and consolidate the oil industry in the US. By 1879, Standard Oil controlled 90 percent of all the oil refined in the US (Yergin, 1991). Had these early oil fields yielded heavier and therefore less marketable crude Standard Oil would likely not have experienced the same rate of investment returns, which would have at least slowed its movement toward consolidation. Consequently, when oil fields began to yield heavier crudes by the 1880s the oil industry (read: Standard Oil) had sufficient capital to produce a market for these previously useless residuals.

By the early 1890s, even with steady improvements to refining technology, crude coming from newly discovered oil fields in Ohio, Indiana, Texas, and California were yielding less than 50 percent light products. Because of an initial lack of demand, the heavier petroleum materials were simply discarded as waste (much was burned in open pits) (Jones, 1978). As fields began yielding higher ratios of heavier crude, however, the oil industry started to think up ways to create demand for this byproduct. Standard Oil formed a special department whose sole purpose was to promote these heavier fuels. One strategy, which proved effective in hindsight, involved giving away—or selling well below cost—thousands of barrels of heavy fuel oil. Another effective strategy involved providing free oil burning appliances to demonstrate the practicality of this fuel (Jones, 1978).

Having monopolized the market, Standard Oil Trust was able to focus a tremendous amount of resources on resolving the "problem" of oil fuel. Had the industry still been in its infancy or had it been composed of many competing firms it is difficult to see how the same level of resources—and a coordination of research and development—could have been directed at turning oil fuel into a desirable commodity in just a couple of years. Yet, in only five years, Standard Oil did just that. While fuel oil represented only 8.6 percent of refined petroleum product output in the US in 1904 that percentage had grown to 40.5 percent by 1909 (Schurr and Netschert, 1960).

Fuel oil's main competition was coal, not alcohol. Oil at this time was beginning to replace coal as fuel for boats (e.g., Britain's Royal Navy began this conversion process in the second decade of the twentieth century [Jones, 1978]), trains, and industrial furnaces (North, 1911). Nevertheless, the development of uses for the heavier elements of crude is important to the story of alcohol. By the turn of the century, it was estimated that gasoline constituted only two percent of raw crude (Baskerville, 1906). By finding a market for the remaining 98 percent of its product, the oil industry was able to increase its influence over the entire fuel market. Granted, the distribution and storage needs differed across petroleum-based fuels. Yet revenues generated across oil fuel types could be pooled for buying more refineries, railroads, pipelines, and oil fields. This is exactly what Standard Oil did, and it brought the industry tremendous gains in efficiency and profitability (Bringhurst, 1979).

Until major oil field discoveries in states like Oklahoma and Texas in the 1930s, the question was not *if* oil would run out but *when*. The title of an article in the magazine *Scientific American* in 1919 asks a question that was on the tips of many tongues: "How long will the oil last?" Proclamations similar to the following predicted the inevitable:

> "In 1914 the world's output of crude oil amounted to 57 million tons and the highest yield of petrol from the whole quantity is placed by Professor Lewes at 1,700,000,000 gallons, of which amount the United States, alone, last year used 1,200,000,000 gallons and Great Britain over 200,000,000." (Hamlin 1915: 631)

The finite nature of oil reserves had been a persistent problem in the minds of many in the U.S. ever since crude was discovered in Pennsylvania in the mid 1800s. In 1919, for example, government geologists estimated U.S. petroleum reserves at 6.74 billion barrels; enough oil for approximately 17 years if consumption remained constant (White, 1920). By 1920, oil had reached a record price of $3.50 a barrel (Dennis, 1985). Even GM for a time jumped on the alcohol bandwagon. From the perspective of automobile manufactures, future car sales would be greatly hampered if the fuel to power these machines was to run out. Making his case in *The Journal of Industrial and Engineering Chemistry*, in an article titled "Motor Fuel from Vegetation," Boyd (1921) presents two figures. These illustrations—Figures 2.1 and 2.2—gave a bleak outlook if alternatives to oil were not invested in quickly.

In 1926, at the 72nd meeting of the American Chemical Society, scientists debated whether "motor-mad America" would be able supply fuel for all automobiles once the "saturation point" was reached, which was estimated at "one motor car for every four persons in the United States" (Science News Letter 1926: 2). The early 1920s signified the high-water mark for public concern about oil's finite quality. Following World War I, the price of crude spiked sharply due to supply shortages (see Table 2.1). Moreover, as illustrated in Table 1, U.S. imports of oil after the war had increased—compared to 1915 levels—at an alarming rate.

THE PRESENT MOTOR FUEL SITUATION

The number of automobiles and trucks in use in the

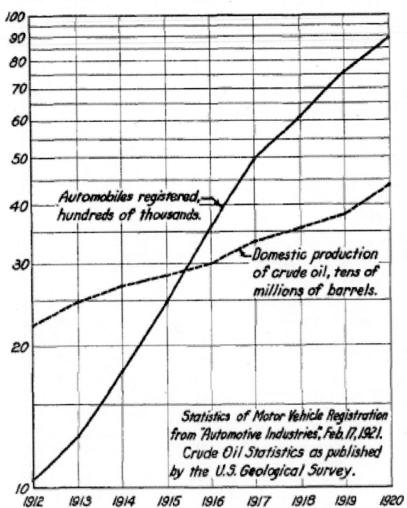

Source: Boyd, T.A. 1921. "Motor Fuel from Vegetation," *The Journal of Industrial and Engineering Chemistry* 13(9): 836.

Figure 2.1. Relation between automobiles registered and production of crude oil.

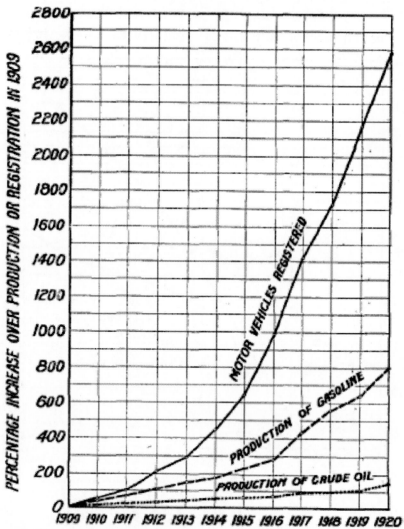

Source: Boyd, T.A. 1921. "Motor Fuel from Vegetation," *The Journal of Industrial and Engineering Chemistry* 13(9): 836.

Figure 2.2. Percentage increases in Motor Vehicles Registered. Production of Crude Oil, and Production of Gasoline. Compared with 1909. [1]. U.S. Bureau of Mines, Bullettin 191, 19: Automobile Industries (Feb. 17, 1921), 306; recent statistics, U.S. Bureau of Mines and Geological Survey.

**Table 2.1. Characteristics of the U.S. Petroleum Market 1890-1935
(in thousands 42 gal. barrels)**

Year	Production	Imports	Exports	Reserves
1890	45,824	0	2,299	0
1895	52,892	0	2,650	0
1900	63,621	0	3,290	2,900,000
1905	134,717	0	3,004	3,800,000
1910	209,557	0	4,288	4,500,000
1915	281,104	18,140	3,768	5,500,000
1920	442,929	106,175	9,295	7,200,000
1925	765,743	61,824	13,337	8,500,000
1930	898,011	62,129	23,705	13,600,000
1935	996,596	32,239	51,430	12,400,000

Source: Series M 138-142 Crude Petroleum—Production, Value, Foreign Trade and Proved Reserves 1859-1970. *Historical Statistics of the United States.*

Alcohol proponents also had, for a time at least, favorable national sentiments working for them. During the Great War, the world witnessed firsthand the disruptive affects that large-scale war had on global petroleum supply chains. Unlike Europe, the U.S. in the 1920s remained decades away from becoming a net importer of oil (this did not occur until 1947 [Terzian, 1991]). But at the time the outlook of oil reserves appeared bleak. For example, while the U.S. only imported 12 percent of its petroleum in 1920 estimates at the time were predicting this figure to rise to nearly 25 percent by 1925 (White, 1920). Like today, issues of energy dependency resonated loudly among the public and politicians alike. As one alcohol proponent proclaimed: "They say we have foreign oil. Well, how are we going to get it in case of war? It is in Venezuela, it is out east, in Persia, and it is in Russia. Do you think that is much defense for your children" (Garvan, 1936: 11)?

The beginning of the Farm Chemurgic Movement has been traced to two articles published in October 1926 (Beeman, 1994; Wright, 1995). One was written by William "Billy" Hale, the then-Director of Organic Chemical Research at the Dow Chemical Company. Appearing in Henry Ford's newspaper *The Dearborn Independent*, Hale's piece, titled "Farming Must Become a Chemical Industry: Development of Co-Products will Solve the Present Agricultural Crisis," made the case for how industry could be effectively built around domestic agricultural commodities. The other article was written by Wheeler McMillen, who at the time was Chair of the Chemistry and Chemical Technology Division of the National Research Council. Published in *Farm and Fireside*, this article, titled "Do We Need This Foundation", similarly called for a marriage of agriculture and industry. The authors' prominence meant both pieces were widely read (Wright, 1993).

The term "chemurgy" combines the Egyptian root for chemistry and the Greek root for work (Wright, 1993). First coined in 1934 by William Hale (1934), the term was intended to evoke a comparison with metallurgy (both were concerned with extracting properties from nature and converting them for industrial ends) (Effland, 1995). To put it simply: the goal of the Farm Chemurgic Movement was to find industrial applications for the growing surplus of farm products following World War I. As H. E. Barnard, a Farm Chemurgic Council member, explained in the *Journal of Farm Economics* in 1938:

> "While our chemists and experiment stations have been showing the farmer how to produce more and more they have given less attention to the importance of developing uses for his [or her] crops outside the satisfaction of the appetites of men [sic] and animals. [...] By using the power produced by gasoline instead of by corn and hay burning horses, we have deprived the farmer of a market for the crops from many million acres." (Barnard, 1938: 119)

One of the reasons behind the movement's emergence was the end World War I. During the Great War, Europe's grain production was severely disrupted, causing them to turn to the US for grain. This sent the price of grain skyward (Santos, 2006). Moreover, once the US entered the war the government began guaranteeing farmers a minimum price of at least $2 a bushel for wheat in an attempt to encourage production. From 1918 to 1919, farm income therefore exceeded nonfarm income by 50 percent (Shideler, 1976). There would have been little incentive for farmers during this time to invest in the Farm Chemurgic Movement and little incentive for industry to look to agriculture for its raw materials.

Following the Armistice in November 1918 farm prices began to drop with the fall harvest. Europe was soon raising crops again, creating a downward pressure on prices. Farmers during this period were also beginning to mechanize. The replacement of human and animal power with technology allowed farmers to cultivate far larger tracts of land than ever before (it has been estimated that farmers placed into production an additional 20 million acres during this period) (Shideler, 1976). Together, these events served to flood an already glutted grain market. Out of this emerged the "farm crisis" that came on the heels of the First World War (Hall, 1973). While commodity prices would soon rebound, concerns over overproduction persisted throughout the 1920s (Shideler, 1976). Improvements in seed varieties, mechanization trends, and advances in artificial fertilizers all contributed to a level of agricultural output that had never been witnessed before. Yet it was not until the Great Depression that conditions became acute enough to cause industry and agriculture to look at each other— through the lens of chemurgy—as solutions to each other's problems.

Grain prices were at near record lows in the early 1930s. In 1933, for example, corn prices slumped to as little as ten cents a bushel (Giebelhaus, 1980a). In this market environment, agricultural commodities were suddenly commercially attractive to those in industry (industry had long been attracted to chemurgic ideas but high market prices kept them from acting on these sentiments). Similarly, those associated with agriculture saw in chemurgy a way to develop new markets for their commodities, which, it was hoped, would result in an upswing in commodity prices. In light of these incentives, industry and agriculture begun to form behind what would eventually be termed the Farm Chemurgic Movement. And within this movement, alcohol was priority number one.

With grain prices at near record lows, Secretary of Agriculture Henry Wallace began to align himself more formally with the chemurgy movement and the chemurgic processes that were being developed at Iowa State College (now Iowa State University) by chemical engineer Orland R. Sweeney (Wallace 1934). As Wallace argued in an editorial in a 1932 issue of *Wallace's Farmer*:

> "Another plan which has been suggested for reaching these men [and women farmers] is to pass a law requiring that all motor fuel burned in this country must contain 10 percent of alcohol made out of corn. [...] If 10 percent of all our motor fuel were alcohol made of corn, the result would be a market for 600,000,000 bushels of corn, or about 22 percent of our ordinary corn crops. [...] Moreover, the day is evidently coming, thirty or forty years hence, when we will have very little petroleum left. If our nation were really intelligent in looking forward, there would be some extensive experimenting done with the making of alcohol out of corn" (Wallace 1932: 11).

The message was timely. Depressed grain prices in the early 1930s had led to a dramatic increase in farm foreclosures throughout the Midwest, which in turn provoked, in a few cases, civil unrest and declarations of martial law (Wright 1993). Grassroots efforts soon emerged throughout the Midwest in an attempt to push the alcohol issue. Chemurgy quickly took on the look of a populist movement.

Several bills emerged out of Midwestern states offering tax incentives to encourage alcohol-gasoline blends. The bills faced stiff resistance by the petroleum industry. To make the legislation palatable to a wider audience, a more moderate bill was proposed for the federal level. This program would offer tax incentives—namely a lower tax rate—to any oil company that agreed to mix a small percentage of alcohol into their gasoline. According to the proposed bill, blends would initially have to contain one percent of alcohol, eventually increasing to a five percent blend. The bill was to take effect on the 1st of July, 1933. The bill had support among members of Congress from agriculture states. Henry Wallace, who helped write the bill, and President Roosevelt publicly supported the bill. The oil industry, in contrast, fiercely opposed any legislation that could potentially erode its market. Enlisting the help of automobile clubs, consumers groups, and others in the motor users' lobby, the oil industry led the charge to defeat the bill, which was in the end successful (Giebelhaus 1980a).

It is worth noting that the nation's automobile clubs, representatives from oil producing states, and elements of the federal government (e.g., Bureau of Standards) had been beneficiaries of major contributions from the American Petroleum Institute throughout the late 1920s and early 1930s (Wright 1993). This involvement is clear from the following memo from April 1933, in which the American Petroleum Institute calls upon its members to act:

> "The situation is so critical that action must be taken at once. Make contact immediately with automobile clubs, motor transport interests, and similar organizations in your territory. Acquaint them with the real nature of their scheme and enlist their support in opposing this legislation" (as quoted in Giebelhaus 1980a: 177-8).

The deep pockets of Big Oil helped a number of times during the 1930s to counter alcohol's populist push from below. It has also been reported that Standard Oil spent in excess of $100,000 to defeat pro-alcohol legislation in Iowa, Nebraska, and South Dakota in 1933 (Giebelhaus, 1979). In a paper presented at the 1939 Farm Chemurgic Council meeting in Omaha, NE, Standard Oil was criticized for threatening to close down their district office in Sioux City, IA if the city's Chamber of Commerce continued in their support of alcohol blend legislation (Standard Oil, in response, claimed it had long planned to close this office) (Giebelhaus, 1979: 50). The same paper also claimed that "somebody" called city business owners reminding them of how many of their customers were employed by major oil companies (Giebelhaus, 1979: 50). A more blatant display of the oil industry's influence occurred when Ethyl Gasoline Corporation (owned by General Motors, Standard Oil of New Jersey, and DuPont) denied licenses to wholesale fuel dealers who sold alcohol blends. This maneuver forced fuel dealers to sell only leaded gasoline or face elimination from the market due to having their supplies disrupted (see e.g., Harbeson, 1940; *Yale Law Journal*, 1939). This practice continued until 1937, when the Department of Justice stepped in to stop it. Eventually the Supreme Court heard the case in 1940, ruling to uphold the actions of the government (Kovarik, 1998).

Chemurgists also had to contend with a physical trait of alcohol that was readily exploited by the oil industry. Alcohol, unless free of water, does not mix with gasoline—what is known as "phase separation" (Hansen, Zhang, and Lyne 2005). The problem of phase separation was repeatedly cited by the oil industry when publically condemning alcohol. During a brief period in 1923 Standard Oil attempted to market an alcohol blend in the Baltimore area (Robert 1983). This was in part a response to the higher price of oil but more importantly the company had access to cheap surplus alcohol from the World War I munitions industry (Bernton et al. 1982). Standard Oil quickly abandoned this experiment, however, noting mechanical problems that were attributed to alcohol's instability in the presence of water (Giebelhaus 1980b; Robert 1983). It was later discovered that Standard Oil did not properly clean its fuel storage tanks. Thus, the problem was not with the fuel *per se* but in how the fuel was handled; though this did not keep the American Petroleum Institute from using this test case as "scientific" justification for opposing alcohol blends (Kovarik 1998).

Another account describes how the problem of phase separation was literally taken on the road by gasoline proponents to discredit alcohol. As described in an early book on chemurgy:

> "Among the neatest tricks to discourage the use of the blend were the repeatedly reported demonstrations with which traveling 'experts' *proved* that alcohol and gasoline do not mix. The self-styled 'experts' conducted this little trick by driving into fill stations and showing proprietors and by-standers that Agrol fluid [alcohol fuel] and gasoline separate into layers. It worked beautifully because the 'experts' used small glass tubes which they carefully washed beforehand. Since a drop of water in a small vial is large enough to cause separation in the small amount of blended fuel in the tube, the demonstration was very effective …". (Borth, 1943: 170, emphasis in original)

The National Farm Chemurgic Council was established in 1935, emerging out of a meeting of industrialists, agricultural leaders, railroad officials, politicians, and scientists. Their first meeting, called the Dearborn Conference, was sponsored by Henry Ford and held in Dearborn, Michigan (Chemical Foundation 1935). That same year the Council requested USDA support to build a large alcohol plant that would use Jerusalem artichokes as its source material and smaller facilities designed for sugarcane and potatoes. With no response from the USDA the Council turned to the Chemical Foundation (an early supporter of chemurgy principles) for funds. Their request was granted. In May, 1936, at the second annual Dearborn Conference, it was announced that the alcohol plant would be located in Atchison, Kansas and would produce 10,000 gallons of alcohol and 30 tons of protein feed a day from Jerusalem artichokes, sweet potatoes, sorghum, corn, potatoes, and grain (Chemical Foundation and Farm Chemurgic Council 1936; Giebelhaus 1980a).

Under the corporate name of the Atchison Agrol Company, the plant went into full commercial operations in March 1937. By the spring of 1938, when the plant was at the peak of its production, Agrol fuel sold in more than 2,000 gas stations in eight Midwestern states. But in April of 1938 the Chemical Foundation withdrew its financial support. There were a variety of reasons for this. The Chemical Foundation was founded during the First World War when the U.S. confiscated a number of German chemical patents. By the late 1930s many of these patents had expired, thus causing a significant revenue stream of the Foundation to dry up. Another problem was that Agrol had locked in contract prices with farmers just before market prices declined (Giebelhaus, 1980a). Plant operators also experienced difficulty obtaining the necessary permits from the Federal Alcohol Tax Unit. The plant's technical director estimated that the cost of compliance with all government regulations was approximately $100,000 (Giebelhaus, 1979).

Chemurgists were also not doing the movement any favors when they selected non-traditional commodities to produce ethyl alcohol. While possessing a more exploitable carbohydrate source than corn, given the technology at the time (Thaysen, et al. 1929), Jerusalem artichokes and sweet potatoes were not part of most farms' commodity profile. Tables 2.2, 2.3, and 2.4 provide a visual sketch of what I mean by this. For the states of Iowa, Illinois, and Missouri—three states that had housed considerable pro-alcohol forces—these tables document the number of commodities produced for sale in at least one percent of all farms for various years between 1920 and 2002 (commodities in bold represent those produced for sale in at least 50 percent of all farms). These tables indicate a strategic misstep among chemurgists in selecting Jerusalem artichokes and sweet potatoes, two commodities, even back when farms still possessed a diverse commodity profile, that were not part of the rural landscape.

This apparent disregard for the cultural and institutional realities of agriculture eventually turned many away from the promises of chemurgy. Secretary of Agriculture Henry Wallace, although an earlier supporter of chemurgy, eventually distanced himself from the movement. In his own words: '[E]nthusiasts tend to burn out the brake bands of their imagination on the subject of farm chemurgy. "Imagineering" has an important place but let us be hard-headed and not count chickens before the eggs are laid' (as quoted in Beeman, 1994: 41).

Nevertheless, alcohol blends had significant popular support within agriculture-dependent states. A survey conducted in 1933 (well before the Atchison Agrol Company was built), of 1,327 users of alcohol blends from Midwest states, shows a consumer willingness to pay a premium for this fuel. The survey was summarized in a 1933 issue of *Science News Letter*:

> "[A survey asking alcohol fuel consumers to give] their impression of alcoholized gasoline on the scores of starting, acceleration, smoothness of operation, anti-knock, power and general motor performance, shows that over 1,100 of them considered the new fuel better than 'straight' non-premium gasoline on starting, and that on all the other points over 1,200 agreed that the new fuel was superior. Most of the dissenting votes merely reported 'no difference noticed'; hardly anyone considered the unblended gasoline better. … [Respondents would] be willing to pay a premium of two or three cents a gallon for alcoholized gasoline, provided the alcohol is produced from surplus grain, the report stated." (Whelpton, 1933: 301)

When the plant in Atchison, Kansas became operational gas stations reported selling alcohol blends for between one and two cents more than conventional gasoline (*Science News Letter*, 1936). Even at this slightly higher price, sales of alcohol blends were said to have been brisk (Berton et al., 1982). After Atchison Agrol Company closed it doors the National Farm Chemurgic Council tried repeatedly to get financial backing from the USDA to build yet another alcohol plant. Again, the government expressed no interest. New Deal policies reflected a shift in strategy, from finding new markets for surplus crops to farm set-asides and crop reduction programs.

Table 2.2. Number of commodities produced for sale in at least one percent of all Iowa farms for various years from 1920 to 2002

1920	(%)	1935	(%)	1915	(%)	1951	(%)	1961	(%)	1978	(%)	1987	(%)	1997	(%)	2002	(%)
Horses	(95)	Cattle	(95)	Cattle	(94)	Corn	(92)	Corn	(91)	Corn	(87)	Corn	(90)	Corn	(79)	Corn	(58)
Cattle	(95)	Horse	(95)	Chicken	(93)	Cattle	(91)	Cattle	(89)	Soybeans	(81)	Soybeans	(68)	Soybeans	(65)	Soybeans	(54)
Chicken	(95)	Chicken	(95)	Corn	(93)	Oats	(91)	Hogs	(83)	Cattle	(69)	Cattle	(60)	Hay	(47)	Hay	(37)
Corn	(94)	Corn	(94)	Horses	(96)	Chicken	(84)	Hay	(82)	Hay	(62)	Hay	(56)	Cattle	(46)	Cattle	(35)
Hogs	(89)	Hogs	(89)	Hogs	(83)	Hogs	(81)	Soybeans	(79)	Hogs	(57)	Hogs	(50)	Hogs	(35)	Horses	(13)
Apples	(84)	Hay	(84)	Hay	(82)	Hay	(80)	Oats	(72)	Oats	(57)	Oats	(44)	Oats	(25)	Hogs	(11)
Hay	(82)	Potatoes	(82)	Oats	(64)	Horses	(74)	Chicken	(42)	Horses	(48)	Horses	(13)	Horses	(10)	Oats	(08)
Oats	(81)	Apples	(81)	Apples	(56)	Soybeans	(41)	Horses	(37)	Chicken	(26)	Sheep	(09)	Sheep	(08)	Sheep	(04)
Potatoes	(62)	Oats	(62)	Soybeans	(52)	Potatoes	(40)	Sheep	(18)	Sheep	(17)	Chicken	(08)	Chicken	(05)	Chicken	(02)
Cherries	(57)	Cherries	(57)	Grapes	(34)	Sheep	(23)	Potatoes	(16)	Wheat	(06)	Ducks	(01)	Goats	(01)		
Wheat	(36)	Grapes	(36)	Potatoes	(28)	Ducks	(23)	Wheat	(05)	Goats	(03)	Goats	(01)				
Plums	(29)	Plums	(29)	Cherries	(25)	Apples	(20)	Sorghum	(05)	Ducks	(02)	Wheat	(01)				
Grapes	(28)	Sheep	(28)	Peaches	(21)	Peaches	(16)	Apples	(04)								
Ducks	(18)	Peaches	(18)	Sheep	(16)	Goats	(16)	Ducks	(03)								
Geese	(18)	Pears	(18)	Plums	(16)	Grapes	(15)	Grapes	(03)								
Stwberry	(17)	Mules	(17)	Pears	(13)	Pears	(13)	Goats	(03)								
Pears	(17)	Ducks	(17)	Rdclover	(12)	Plums	(10)	Geese	(03)								
Mules	(14)	Wheat	(14)	Mules	(12)	Wheat	(06)										
Sheep	(14)	Geese	(14)	Geese	(11)	Rdclover	(06)										
Timothy	(10)	Sorghum	(10)	Stwberry	(09)	Geese	(04)										
Peaches	(09)	Barley	(09)	Ducks	(09)	Popcorn	(04)										
Bees	(09)	Rdclover	(09)	Wheat	(09)	Timothy	(03)										
Barley	(09)	Stwberry	(09)	Timothy	(08)	Swtpotato	(02)										
Raspby	(07)	Soybeans	(07)	Geese	(08)	Swtcorn	(02)										
Turkeys	(07)	Raspby	(07)	Rye	(08)	Turkeys	(02)										
Wtmelon	(06)	Bees	(05)	Popcorn	(06)												
Sorghum	(06)	Timothy	(05)	Swt corn	(05)												
Goosebry	(03)	Turkeys	(04)	Rasbry	(05)												
Swt corn	(02)	Rye	(02)	Bees	(04)												
Apricots	(02)	Popcorn	(02)	Sorghum	(02)												
Tomatoes	(02)	Swt corn	(02)														
Cabbage	(01)	Goats	(01)														
Popcorn	(01)																
Currents	(01)																
n – 34		n – 33		n – 29		n – 26		n – 17		n – 12		n – 12		n – 10		n – 9	

Source: US Census of Agriculture, 1920-2002

Prepared by Michael S. Carolan, PhD; Department of Sociology; Colorado State University; Fort Collins, CO; 80526; mcarolan@colostate.edu.

Table 2.3. Number of commodities produced for sale in at least one percent of all Illinois farms for various years from 1920 to 2002

1920	(%)	1935	(%)	1945	(%)	1954	(%)	1964	(%)	1978	(%)	1987	(%)	1997	(%)	2002	(%)
Horses	(95)	Chicken	(93)	Chicken	(89)	Corn	(89)	Corn	(84)	Corn	(75)	Corn	(75)	Corn	(66)	Soybeans	(57)
Chicken	(95)	Cattle	(91)	Cattle	(87)	Cattle	(82)	Cattle	(66)	Soybeans	(70)	Soybeans	(70)	Soybeans	(69)	Corn	(56)
Cattle	(92)	Corn	(81)	Corn	(83)	Chicken	(70)	Soybeans	(60)	Cattle	(39)	Cattle	(39)	Cattle	(38)	Hay	(38)
Corn	(87)	Horses	(80)	Hogs	(70)	Hogs	(61)	Hogs	(50)	Hay	(37)	Hay	(37)	Hay	(35)	Cattle	(37)
Hogs	(84)	Hogs	(73)	Horses	(67)	Oats	(50)	Wheat	(46)	Hogs	(22)	Wheat	(22)	Wheat	(24)	Horses	(13)
Hay	(68)	Potatoes	(55)	Oats	(53)	Soybeans	(50)	Hay	(37)	Wheat	(22)	Hogs	(21)	Hogs	(20)	Wheat	(10)
Apples	(68)	Apples	(47)	Hay	(44)	Hay	(38)	Chicken	(33)	Horses	(12)	Horses	(13)	Horses	(11)	Hogs	(05)
Oats	(67)	Hay	(44)	Soybeans	(42)	Horses	(37)	Oats	(33)	Oats	(11)	Oats	(11)	Oats	(10)	Oats	(03)
Potatoes	(61)	Oats	(36)	Apples	(31)	Sheep	(18)	Horses	(21)	Chickens	(07)	Chickens	(05)	Chickens	(05)	Sheep	(03)
Cherries	(50)	Grapes	(33)	Wheat	(30)	Potatoes	(11)	Sheep	(15)	Sheep	(03)	Sorghum	(02)	Sorghum	(02)	Chickens	(02)
Wheat	(48)	Cherries	(32)	Cherries	(24)	Rye	(05)	Kdclover	(12)	Ducks	(01)	Goats	(01)	Goats	(01)		
Peaches	(44)	Wheat	(28)	Grapes	(23)	Apples	(05)	Potatoes	(04)	Apples	(01)						
Grapes	(32)	Pears	(26)	Pears	(22)	Swtpatoe	(05)	Rye	(02)	Swt corn	(01)						
Pears	(32)	Plums	(23)	Mules	(21)	Kdclover	(03)	Swtpatoe	(02)	Sorghum	(01)						
Plums	(40)	Mules	(20)	Potatoes	(18)	Peaches	(03)	Mules	(02)								
Mules	(26)	Sheep	(15)	Rdclover	(13)	Cherries	(04)	Sorghum	(01)								
Ducks	(21)	Soybeans	(14)	Sheep	(13)	Pears	(03)	Barley	(01)								
Geese	(15)	Swtpatoe	(13)	Plums	(12)	Grapes	(03)	Swt corn	(01)								
Bees	(12)	Ducks	(13)	Swtpatoe	(05)	Barley	(05)	Apples	(01)								
Sheep	(11)	Rdclover	(08)	Ducks	(05)	Ducks	(03)	Peaches	(01)								
Swtpatoe (11)		Geese	(08)	Geese	(03)	Plums	(03)	Ducks	(01)								
Strwbcry(11)		Bees	(08)	Turkeys	(04)	Swt corn	(02)										
Rye	(08)	Turkeys	(04)	Bees	(03)	Tomatoes	(01)										
Barley	(07)	Cowpeas	(04)	Swt corn	(03)	Popcorn	(01)										
Rdclover (07)		Rasphry	(04)	Strwberry	(03)	Hen	(01)										
Sunghum (07)		Tomatoes	(03)	Apricots	(03)	Turkeys	(01)										
Cowpeas (07)		Swt corn	(03)	Goats	(03)	Geese	(01)										
Dikiberry (06)		Peaches	(03)	Popcorn	(02)												
Turkeys	(03)	Sorghum	(02)	Tomatoes	(02)												
Timothy	(03)	Barley	(02)	Kingley	(02)												
Califower (02)		Kindug	(02)	Cowpeas	(02)												
Swt corn	(02)	Popcorn	(02)	Barley	(02)												
Onions	(02)	Snapbean	(02)	Rye	(02)												
Tomatoe	(02)	Strwberry	(02)	Redtop	(02)												
Grn bean	(01)	Dlk berry	(02)	Lespedez	(01)												
Cucumbs	(01)	Goats	(01)	Swtclover	(01)												
		Timothy	(01)														
		Swtclover	(01)														
		Watermel	(01)														
		Cabbage	(01)														
n = 36		n = 41		n = 37		n = 28		n = 21		n = 14		n = 11		n = 11		n = 10	

Source: US Census of Agriculture, 1920-2002.
Prepared by Michael S. Carolan, PhD; Department of Sociology; Colorado State University; Fort Collins, CO; 80526; mcarolan@colostate.edu.

Table 2.4. Number of commodities produced for sale in at least one percent of all Missouri farms for various years from 1920 to 2002

1920	(%)	1935	(%)	1945	(%)	1954	(%)	1964	(%)	1978	(%)	1987	(%)	1997	(%)	2002	(%)
Chickens	(95)	Chickens	(90)	Cattle	(88)	Cattle	(88)	Cattle	(87)	Cattle	(84)	Cattle	(73)	Cattle	(69)	Cattle	(61)
Cattle	(92)	Cattle	(89)	Chickens	(88)	Chickens	(88)	Hay	(76)	Hay	(55)	Hay	(61)	Hay	(57)	Hay	(54)
Horses	(89)	Corn	(72)	Corn	(67)	Corn	(67)	Corn	(58)	Soybeans	(47)	Soybeans	(36)	Soybeans	(37)	Horses	(22)
Corn	(84)	Hogs	(67)	Horses	(65)	Horses	(65)	Hogs	(55)	Hogs	(13)	Corn	(28)	Corn	(17)	Soybeans	(20)
Hogs	(83)	Horses	(66)	Hogs	(61)	Hogs	(61)	Chickens	(54)	Corn	(12)	Horses	(28)	Horses	(17)	Corn	(15)
Apples	(76)	Hay	(51)	Hay	(50)	Hay	(50)	Wheat	(52)	Horses	(31)	Hogs	(19)	Wheat	(14)	Wheat	(08)
Peaches	(68)	Potatoes	(50)	Potatoes	(40)	Potatoes	(40)	Soybeans	(39)	Wheat	(30)	Wheat	(15)	Hogs	(12)	Chickens	(04)
Potatoes	(61)	Apples	(12)	Peaches	(39)	Oats	(39)	Horses	(32)	Sorghum	(23)	Sorghum	(12)	Chickens	(09)	Hogs	(03)
Cherries	(57)	Peaches	(10)	Apples	(31)	Wheat	(31)	Potatoes	(25)	Chickens	(17)	Chickens	(11)	Sorghum	(07)	Sheep	(02)
Wheat	(54)	Mules	(35)	Mules	(25)	Soybeans	(25)	Oats	(20)	Feecue	(13)	Sorghum	(07)	Sheep	(03)	Sorghum	(02)
Hay	(51)	Grape	(29)	Grapes	(25)	Mules	(25)	Sorghum	(15)	Sheep	(10)	Feecue	(03)	Goats	(03)	Oats	(01)
Mules	(35)	Sorghum	(24)	Cherries	(24)	Sorghum	(24)	Mules	(13)	Ducks	(07)	Oats	(03)	Oats	(01)		
Oats	(17)	Cherries	(21)	Oats	(24)	Swpatoe	(21)	Sheep	(13)	Oats	(06)	Sheep	(02)				
Plums	(15)	Oats	(21)	Peas	(21)	Sheep	(21)	Swpatoe	(10)	Cotton	(05)	Goats	(02)				
Pecans	(32)	Pecans	(23)	Wheat	(19)	Peaches	(19)	Mules	(10)	Apples	(05)	Bees	(01)				
Grapes	(23)	Wheat	(22)	Plums	(18)	Apples	(18)	Cotton	(10)	Turkeys	(05)	Cotton	(01)				
Turkeys	(22)	Plums	(22)	Sheep	(17)	Barley	(17)	Ducks	(08)	Mules	(03)	Ducks	(01)				
Swpatoe	(15)	Sheep	(22)	Swpatoe	(11)	Cotton	(14)	Rdclover	(08)	Peaches	(03)	Mules	(01)				
Sorghum	(15)	Swpatoe	(18)	Sorghum	(14)	Bees	(08)	Peache	(07)			Rdclover	(01)				
Sheep	(14)	Oats	(18)	Lespedez	(08)	Peas	(08)	Apples	(06)								
SwtSough	(12)	Bees	(11)	Soybeans	(08)	Grapes	(07)	Barley	(05)								
Bees	(11)	Ducks	(08)	Cotton	(07)	Cherries	(05)	Timothy	(05)								
Geese	(10)	Geese	(08)	Bees	(05)	Plums	(05)	Lespedez	(05)								
Ducks	(09)	Turkeys	(08)	Stwberry	(04)	Lespedez	(05)	Goats	(05)								
Stwbeny	(06)	Cotton	(07)	Rdclover	(04)	Ducks	(04)	Grapes	(03)								
Stwberry	(04)	Soybeans	(06)	Turkeys	(03)	Rye	(03)	Peas	(03)								
Blkberry	(04)	Stwberry	(04)	Goats	(03)	Timothy	(03)	Cherries	(03)								
Rye	(03)	Lespedez	(04)	Ducks	(03)	Goats	(03)	Rye	(03)								
Soybeans	(03)	Goats	(04)	Barley	(03)	Geese	(03)	Geese	(03)								
Cotton	(03)	Rdclover	(03)	Tomatoes	(02)	Rdclover	(02)	Plums	(02)								
Tobacco	(03)	Blkberry	(03)	Timothy	(02)	Turkeys	(01)										
Tomatoes	(03)	Tomatoes	(03)	Rye	(02)												
Goats	(02)	Timothy	(02)	Geese	(02)												
Timothy	(02)	Rye	(02)	Apricots	(01)												
Raspby	(02)	Wmellon	(02)	Blkberry	(01)												
Rdclover	(01)	Snapbean	(01)														
Buncorn	(01)	Raspby	(01)														
Guineas	(01)																
Cabbage	(01)																
Dryonion	(01)																
Watermel	(01)																
Pecans	(01)																
n = 41		n = 35		n = 35		n = 31		n = 29		n = 18		n = 18		n = 12		n = 11	

Source: US Census of Agriculture, 1920-2002.

Prepared by Michael S. Carolan, PhD; Department of Sociology; Colorado State University; Fort Collins, CO; 80526; mcarolan@colostate.edu.

It is rare in the U.S. that technocratic optimism lands on the losing side of a struggle for the popular imagination, particularly in an era that gave birth to such slogans as "Better Things for Better Living … Through Chemistry" (adopted by DuPont in 1935). Unlike, say, early steam cars, which came to represent "old fashionedness" (Hard and Jamison, 1997: 148), ethyl alcohol rested upon, as an article in *Popular Mechanics* (1933: 488) proclaims, "the miracles of modern chemistry". During the initial decades of the twentieth century, chemurgy represented the forefront of chemistry in the US (Beeman, 1994). But this status was not to last.

Chemists would soon turn their attention to the chemical properties of petroleum. "By the late 1930s", as one historical account of plastics in the US explains, "the push to exploit basic chemical raw materials derived from coal, natural gas, or petroleum' had reached a fevered pitch" (Meikle, 1995: 82). And as chemists turned towards petroleum they turned away from the agro-materials of chemurgy:

> "Acrylic plastic was not seriously employed until the Second War provided a need for strong, lightweight, optically perfect thermoplastic sheets that could be formed into aerodynamic shapes for airplane cockpit covers and gunners' enclosures, but it first appeared on the market during the 1930s. … The acrylics also provided another example of the fundamental shift in raw materials from *cellulose* and coal tar to petroleum and natural gas." (Meikle, 1995: 84, emphasis added)

While ethyl alcohol continued to be necessary for the production of rubber during World War II, the plastics revolution (accelerated by WWII) suddenly made chemurgy look a little too much like its namesake metallurgy—outdated.

But I am getting ahead of myself. There is still more to this story. In the next chapter, I detail additional forces that had the cumulative effect of leading to alcohol's disappearance from the fuel market by 1940.

REFERENCES

Barnard, H. E. (1938) 'Prospects for Industrial Uses for Farm Products', *Journal of Farm Economics* 20(1): 119-33.

Baskerville, Charles (1906) 'Free Alcohol in the Arts and as Fuel', in Albert Shaw (ed.), *The American Monthly Review of Reviews* (New York: The Review of Reviews Company): 211-14.

Beeman, Randal (1994) 'Chemivisions: The Forgotten Promise of the Chemurgy Movement', *Agricultural History* 68(4): 23-45.

Berton, Hal, William Kovarik & Scott Sklar 1982. *The Forbidden Fuel: Power Alcohol in the Twentieth Century* New York: Boyd Griffin.

Bolles, Albert 1878. *Industrial history of the United States*. Norwich, CT: Henry Bill Publishing.

Borth, Christy (1943) *Pioneer of Plenty: Modern Chemists and their Work* (New York: The New Home Library).

Bringhurst, Bruce (1979) *Antitrust and the Oil Monopoly: The Standard Oil Cases, 1890-1911* (Westport, CN: Greenwood Press).

Chemical Foundation 1935. *Proceedings of the Dearborn Conference of Agriculture, Industry, and Science* New York: Chemical Foundation.

Chemical Foundation and Farm Chemurgic Council (1936) *Proceedings of the Second Dearborn Conference of Agriculture, Industry, and Science* (Dearborn, MI: Farm Chemurgic Council, Inc. and New York: Chemical Foundation).

Dennis, Michael (1985) 'Drilling for Dollars: The Making of US Petroleum Reserve Estimates, 1921-25', *Social Studies of Science* 15(2): 241-65.

Effland, Anne (1995) 'New Riches from the Soil: The Chemurgic Ideas of Wheller McMillen', *Agricultural History* 69(2): 288-97.

Garvan, Francis (1936) 'Introduction', in F. Garvan (ed.), *Proceedings of the Second Annual Dearborn Conference of Agriculture, Industry, and Science* (New York: The Chemical Foundation): 2-8.

Giebelhaus, August (1979) 'Resistance to Long-Term Energy Transition: The Case of Power Alcohol in the 1930s', in Lewis Perelman, August Giebelhaus & Michael Yokell (eds), *Energy Transitions: Long-Term Perspectives* (Boulder, CO: Westview Press): 35-63.

Giebelhaus, August (1980a) 'Farming for Fuel: The Alcohol Motor Fuel Movement of the 30s', *Agricultural History* 54(1): 173-84.

Giebelhaus, August (1980b) *Business and Government in the Oil Industry: A Case Study of Sun Oil, 1876-1945* (Greenwich, CN: JAI Press).

Gradenwitz, Alfred (1907) 'The Denatured Alcohol Situation', *Scientific American Supplement* S1 (1688): 393.

Hale, William (1934) *The Farm Chemurgic: Forward the Star of Destiny Lights Our Way* (Boston, MA: Stratford).

Hall, Tom 1973. "Wilson and the food crisis: Agricultural prices controls during World War I," *Agricultural History* 47(1): 24-46.

Hamlin, M. 1915. 'Liquid Fuel for Internal Combustion Engines', *The Journal of Industrial and Engineering Chemistry* 7(7): 631-32.

Hansen, Alan, Qin Zhang, and Peter Lyne 2005. "Ethanol—diesel fuel blends—a review," *Bioresource Technology* 96(3): 277-285.

Harbeson, Robert (1940) 'The Present Status of the Sherman Act', *Michigan Law Review* 39(2): 189-212.

Hordeski, Michael (2007) *Alternative Fuels: The Future of Hydrogen* (Lilburn, GA: The Fairmont Press).

Jones, Gareth 1978. 'The Oil Fuel Market in Britain 1900-14—A Lost Cause Revisited', *Business History* 20(2): 131-52.

Kovarik, Bill 1998. 'Henry Ford, Charles F. Kettering and the Fuel of the Future', *Automotive History Review* 32: 7-27.

Kovarik, Bill (2003) 'Ethyl: The 1920s Environmental Conflict over Leaded Gasoline and Alternative Fuels', Paper to the American Society for Environmental History, Annual Conference (26-30 March), Providence, RI. Available at: http://www.runet.edu/~wkovarik/papers/ethylconflict.html, last accessed December 7, 2007.

Long, Clarence. 1960. *Wage and Earning in the United States, 1860-1890.* Princeton, NJ: Ayerpub.

Long, J. H. 1906. 'The Question of Tax-Free Alcohol', *Science* 23(580): 234-35.

McCarthy, Tom (2001) 'The Coming Wonder? Foresight and Early Concerns about the Automobile', *Environmental History* 6(1): 46-74.

North, Sydney (1911) *Oil Fuel: Its Supply, Composition and Application* (London: Griffin).

Popular Mechanics 1912. Prize for Alcohol Fuel Tests, October: 565.

Popular Mechanics (1933) 'Miracles from Test Tubes', *Popular Mechanics* 60 (October): 488-92.

Hard Mikael & Andrew Jamison (1997) 'Alternative Cars: The Contrasting Stories of Steam and Diesel Automotive Engines', *Technology in Society* 19(2): 145-60.

Herrick, Rufus 1907. *Denatured or Industrial Alcohol* New York: John Wiley and Sons.

Meikle, Jeffrey (1995) *American Plastic: A Cultural History* (New Brunswick, NJ: Rutgers University Press).

Nelson, Derek 1995. *Moonshiners Bootleggers and Rumrunners* Osceola, WI: Motorbooks International.

Robert, Joseph (1983) *Ethyl: A History of the Corporation and the People Who Made It* (Charlottesville, VA: University Press of Virginia).

Santos, Joseph 2006. "Going against the grain: Why did wheat marketing in the United States and Canada evolve so differently?" unpublished manuscript, Department of Economics, South Dakota State University, http://www.thebhc.org/publications/BEHonline/2006/santos.pdf.,last accessed December 7, 2007.

Schurr, Sam & Bruce Netschert (1960) *Energy in the American Economy 1850-1975* (Baltimore, MD: The Johns Hopkins University Press).

Science News Letter 1926. Chemists Believe Future Oil Problems Can Be Met, *The Science New Letter* 9(283): 2-4.

Science News Letter 1928. 'Alcohol and Bitumen Sands for Power', *The Science News Letter* 14(391): 209-10.

Science News Letter 1933. 'Alcohol-Gasoline Mixture Suggested as Motor Fuel', *The Science News Letter* 23(620): 116.

Science News Letter (1936) 'Power-Alcohol Plant Now in Commercial Production', *The Science News Letter* 30(809): 237.

Shideler, James (1976) *Farm Crisis, 1919-1923* (New York: Greenwood Press).

Terzian, Pierre (1991) 'The Gulf Crisis: The Oil Factor—An Interview with Pierre Terzian', *Journal of Palestine Studies* 20(2): 100-05.

Thaysen, Aage, William Bakes & Brian Green (1929) 'On the Nature of the Carbohydrates Found in the Jerusalem Artichoke', *The Biochemical Journal* 23(3): 444-55.

Wallace, Henry. 1934. "The Social Advantages and Disadvantages of the Engineering Scientific Approach to Civilization," *Science* 79 (5 January): 3

Wallace, Henry. 1932. "The Voluntary Allotment Plan," *Wallace's Farmer* 24 (11 November): 11.

Whelpton, P. K. (1933) 'Alcoholized Gasoline Approved in Tests', *Science News Letter* 23: 631: 301-2.

White, David (1920) 'The Petroleum Resources of the World', *Academy of Political and Social Science Annals* 89: 111-34.

Wright, F. B. (1907) *A Practical Handbook on the Distillation of Alcohol from Farm Products, 2nd edition* (New York: Camelot Press).

Wright, David (1995) 'Agricultural Editors Wheeler McMillen and Clifford V. Gregory and the Farm Chemurgic Movement', *Agricultural History* 69(2): 272-87.

Wright, David (1993) 'Alcohol Wrecks a Marriage: The Farm Chemurgic Movement and the USDA in the Alcohol Fuels Campaign in the Spring of 1933', *Agricultural History* 67(1): 36-66.

Yale Law Journal (1939) 'A Possible Extension of the Scope of the Patent Monopoly', *The Yale Law Journal* 48(6): 1089-95.

Yergin, Daniel. (1991) *The Prize: The Epic Quest for Oil Money and Power* (New York: Simon and Schuster).

In: A Sociological Look at Biofuels
M. S. Carolan, pp. 29-43

ISBN: 978-1-60876-708-3
© 2010 Nova Science Publishers, Inc.

Chapter 3

GASOLINE GAINS MOMENTUM

The future of ethyl alcohol looked bright in the years immediately following the repeal of the alcohol sin tax in 1906. The following excerpt, written in 1907, is by an individual who provided expert assistance in drawing up the aforementioned bill. It was not uncommon for scientists and engineers familiar with alcohol's properties to publically extol similar virtues:

> "Alcohol as a motor fuel in internal combustion engines has been shown by experiment to be equally as efficient as gasoline. A pound of alcohol produces as much power as a pound of gasoline. If price alone controlled, alcohol would not displace gasoline as a producer of power unless gasoline should rise in price to 25 cents a gallon. But other considerations, aside from cost, make alcohol as a motor fuel preferable to gasoline. Gasoline is highly inflammable; alcohol is not. Many farmers can not secure insurance with gasoline stored on their premise; there is no such objection to alcohol. Gasoline is also at a disadvantage as a power producer because of the offensive exhaust gases; no disagreeable odors attend the use of alcohol. Alcohol is both safer and more agreeable to use than gasoline. With the price of the two articles only a few cents apart alcohol ought to displace gasoline to an extent as a power producer." (Gates 1907: 61)

But we know that was not to be. Instead, gasoline emerged the victor, while alcohol almost entirely disappeared from the fuel marketplace by 1940.

Sociologists of technology are quick to point out that technological artifacts cannot be reduced to "things". Today's gas-powered automobile, for example, would be considerably less effective were it not for oil, roads, automotive engineers, gas stations, car companies, government fuel taxes, pro-automobile cultural imperatives, and the like. As illustrated in the movie *Back to the Future Part III* (1990), where Marty McFly (played by Michael J. Fox) finds himself in 1885 with a gasoline powered Delorean, cars are nothing more than processed raw materials when abstracted from the system out of which they emerged. When examining technology and technological change sociologists often speak of a "socio-technical system" to emphasize this contextual quality. User practices, the framing of a technological artifact, "efficiencies" created by the technology, and market price, all affect and are affected by the socio-technical system within which a given technological artifact is embedded. Even understandings of "fuel" are shaped by these system conditions. Nothing is inherently fuel. Fuel is simply a term for a carrier of energy. A system must be in place that requires a particular carrier of energy if said carrier is to be called "fuel". Even oil, before society organized so as to give it the label of fuel, was once viewed "with indifference or annoyance" (Bolles 1878: 772).

And this context matters. When thought of in abstract isolation, a technological artifact appears innocuous and highly mutable. Untethered to the sociological realities that affect and are an effect of its existence, technological "things" look as if they could head in whatever direction the winds of consumer or political preference are blowing. In reality, however, technological artifacts over time gather momentum. After thousands of miles of railroad track in the U.S. were laid with a standard gauge (the space between rails) of 4 feet and 8 ½ inches, and thousands of engines and boxcars were designed according to these specifications, the transaction costs of adopting another standard gauge became simply too great to make further change likely. Yet this does not imply technological determinism. The trajectory of the railroad was not predetermined with the locking-in of a 4 feet and 8 ½ inch gauge. A host of contextual factors have given shape to the railroad socio-technical system as we know it today. For example, explaining the variability between French and U.S. rail systems, Berk (1994) notes how the latter has been shaped considerably by a prominent and powerful long-haul trucking industry, important court cases, and the efficiencies created when a society invests heavily in car and truck transportation.

Within the social construction of technology (SCOT) literature, momentum becomes most notable once "closure" and "stability" are reached; points at which interpretative flexibility begins to harden. Drawing from the Duhem-Quine principle in philosophy of science, SCOT scholars make the important observation that self-evident, objective reasons cannot alone be used to explain the developmental trajectory of a technology. As Feenberg (1999: 79) explains while discussing the social construction of economic efficiency:

"Before the efficiency of a process can be measured, both the type and quality of output have to be fixed. Thus economic choices are necessarily secondary to clear definitions of both the problems to which technology is addressed and the solutions it provides. But clarity on these measures is often the outcome rather than the presupposition of technological development. For example, MS DOS lost the competition with Windows graphical interface, but not before the very nature of computing was transformed by a change in the user base and in the types of tasks to which computers were dedicated. A system that was more efficient for programming and accounting tasks proved less than ideal for secretaries and hobbyists interested in the ease of use."

This brings me back to the aforementioned concept of "interpretive flexibility:" the idea that the meanings given to a technology are effects of social processes (Pinch and Bijker 1984). An exemplary work on this subject is MacKenzie's (1990) study on the development of ballistic missile technology, where he provides a detailed account of the social processes that went into constructing an agreed upon definition of what "missile accuracy" meant. Similarly, in their study of early bicycles, Pinch and Bijker (1984; Bijker 1995) note initial competing views of what it meant to produce a "working" bike—that is, was it fast, safe, smooth riding, etc? Yet, while SCOT scholars pay significant analytic attention to the interpretative structures that surround technological artifacts, they too note that once a technology becomes entrenched the "network dynamics of convergence" (Callon 1995: 315) make going back difficult. Closure therefore "results in one artifact—that is, one meaning as attributed by one social group—becoming dominant across all relevant social groups" (Bijker 1995: 271).

Writing on how the design stage of a technology compares to its diffusion stage, Sclove (1995) notes how the former is often more flexible in terms of what is considered "possible." This is not to suggest, however, a state of "anything goes". New technology does not emerge out of thin air (the so-called "black swan" [Taleb 2008] is a chimera). A preexisting order conditions new technologies, even during the design stage. Nevertheless, research consistently shows that technology is often "open" to change during the design stage, before networks begin to stabilize around one particular form (Hamlett 1992; Kim 2008).

As late as the 1920s, chemists and engineers looked favorably upon alcohol as a fuel source. For example, a paper published in *The Journal of Industrial and Engineering Chemistry* in 1920 spoke of how alcohol would inevitably replace gasoline as the primary fuel for internal combustion engines, noting the former's many advantages over the latter: more power, increased mileage, less noxious exhaust, and its ability to eliminate engine knock (Tunison 1920). For a short time, even GM promoted alcohol blends. They looked to alcohol for its anti-knock properties and out of a fear that petroleum reserves would soon run out. The top engineer for GM, Thomas Midgley, drove an automobile with a high compression ratio from Dayton, OH to a 1921 Society of Automotive Engineers meeting in Indianapolis, OH using a 30 percent alcohol fuel blend. At the meeting Midgley proclaimed: "Alcohol has tremendous advantages and minor disadvantages, [the former include] clean burning and freedom from any carbon deposit […] [and] tremendously high compression under which alcohol will operate without knocking" (as quoted in Kovarik 1998:14).

Between 1900 and 1921, over 150 scholarly and popular press style articles had already been published under the heading "Alcohol as Fuel" according to the *Readers Guide to Periodical Literature* (Kovarik 2003). Yet, in a famous book chronicling the history of leaded gasoline—*Ethyl: A History of the Corporation and People Who Made It* (Robert 1983)—alcohol is mentioned nowhere as either an early anti-knocking agent or as a potential substitute for gasoline. In the words of the author, writing on the subject of leaded gasoline's discovery,

> "It is appropriate to reflect for a moment on the significance of the discovery [of the anti-knocking properties of lead] on December 9, 1921. Obviously the innovation was the force that created a profitable business, but this is not the main point. *Tetraethyl lead made possible the early development of the modern high-compression, high-powered internal combustion engine.*" (Robert 1983: 109-10) (my emphasis)

What is remarkable about this revisionist historical account is its failure to acknowledge that a GM engineer—just a couple months prior to this December 1921 discovery—spoke of the "tremendously high compression under which alcohol will operate without knocking." This becomes even more remarkable when it is revealed that the individual who discovered the anti-knock properties of lead in December of 1921 is the same person who earlier that year praised alcohol at the annual Society of Automotive Engineers meetings: Thomas Midgley. After its discovery many suddenly forget about the positive properties of alcohol fuel. For example, while Midgley spoke of alcohol's anti-knock properties as early as 1921 he made the following claim in a 1925 paper presented at the American Chemical Society: "So far as science knows at the present time, tetraethyl lead is the only material available which can bring about these [anti-knock] results […] and unless a grave and inescapable hazard exists in the manufacture of tetraethyl lead, its abandonment cannot be justified" (as quoted in Kovarik 2003: 13). And why was a chemical as pernicious as lead preferred over alcohol as an anti-knock additive? In a word: profit.

The process by which lead was introduced into gasoline was quickly patented by GM (Robert 1983). This effectively secured a monopoly for GM over leaded gasoline. Conversely, alcohol could have been produced by anyone through simple processes and inexpensive technologies. In 1923, GM calculated that lead would allow them to capture 20 percent of the gasoline market resulting in revenues in excess of $36 million a year in the first years of commercial applications. They calculated that profits within a ten year span would reach $360 million a year and by the 1950s profits would be in the billions (Kovarik 2005: 385). GM also understood the self-reinforcing effect higher compression ratio engines would eventually have on the fuel market. As more engines with high compression ratios made their way onto the roads drivers would become increasingly dependent on those who held the patents to leaded gasoline. Holding the patents to leaded gasoline guaranteed that the patent holder would make a profit from every high compression car sold; if not from the vehicle itself, then from later fuel sales (Loeb 1995: 83).

Soon after its discovery, GM, which possessed an interlocking directorship with DuPont Chemical Company, contracted with DuPont and Standard Oil of New Jersey to manufacture tetraethyl lead (Rosner and Markowitz 1985). Leaded gasoline was made available in selected markets on February 1, 1923. In August 1924, GM, DuPont, and Standard Oil pooled their patents for manufacturing tetraethyl lead and created the Ethyl Gasoline Corporation (Standard Oil was included in this venture because at the time they owned the patent for the most efficient process for manufacturing lead) (Robert 1983). And the company possessed a virtual monopoly over the additive until leaded gasoline was phased out in the 1970s.

The previous chapter mentions how the Farm Chemurgic Movement was looking to establish an alcohol industry in the 1930s based upon commodities other than corn, soybeans, and grain. Not mentioned, however, was how this was occurring at the very moment when these commodities were being locked in, which, concomitantly, meant high-starch commodities like artichokes (the commodity of choice among chemurgists) were being locked out. At the time, the USDA was unveiling its various New Deal agricultural programs. The earliest of New Deal agricultural policies were concerned with limiting production. The Agricultural Adjustment Administration (AAA), which came into being as a result of the Agricultural Adjustment Act of 1933, emerged out of the belief that farmers could improve their financial position by employing the same methods used in industry—namely, by adjusting their production to that of demand. In 1936, however, these policies were ruled unconstitutional and were quickly replaced by programs directed toward different ends. Devastating droughts in 1934 and 1936 revealed the need for agricultural policies that helped establish large reserves so as to protect the nation (and the nation's farmers) against agriculturally "lean" years (Saloutos 1974).

By executive order, the Commodity Credit Corporation was established in 1933. This made nonrecourse loans available to corn and cotton growers. The program was widely popular. Total corn placed under loan in 1933-1934, for example, exceeded 271 million bushels (Kramer 1983). The program's popularity in addition to the drought of 1934 prompted the USDA to extend the loan program, thereby forming the basis for what was called an "ever-normal granary:" "a definite system whereby supplies following years of drought or other great calamity would be large enough to take care of the consumer, [and] under which the farmer would not be unduly penalized in years of favorable weather" (Shepherd 1947: 40). The loan program soon merged with crop insurance, and together these programs sought to stabilize grain supplies and prices (Kramer 1983). While crop insurance was initially only offered to wheat growers (and soon thereafter to cotton and flax farmers) the program was expanded in 1944 to include such crops as corn, soybeans, barley, and the like (Kramer 1983). It is against this policy backdrop, which offered incentives to raise corn and soybeans, that we must place the aforementioned actions of the Farm Chemurgic Movement. And it is against this backdrop that we can find some understanding as to why commercial success of non-corn derived alcohol fuels in the 1930s was so elusive.

Couple with this the reality that farms were beginning to mechanize in the 1930s. This is significant because as farmers substituted mechanical power for human and animal power it became difficult for them to alter the commodity profiles of their operations. To put it another way: once a farmer invested in, say, a 12-foot combine for corn there was an immediate economic disincentive (what economics call a transaction cost) for them to switch to Jerusalem artichokes and an economic incentive for them to further specialize in corn (to spread the cost of the investment over as large an operation as possible). An indication of this trend can be seen in how the number of human hours per acre of corn dropped during the first half of the twentieth century: from 38 hrs/acre in 1900 to 33 in 1920-4, 28 in 1930-4, and 25 in 1940-4 (and by 1955-9 the figure had dropped to only 10 hours of labor per acre) (Rasmussen 1962: 583). By the mid 1940s there were, for the first time, more tractors on farms in the United States than horses and mules (Sargen 1979). Similarly, tractor powered combines were beginning to be adopted by a larger segment of the farming population (Anderson 2002). By 1930, approximately 50,000 corn pickers were in operation, which was five times more than the number in use just a decade earlier (Colbert 2002). By the 1930s mechanical corn pickers became the most cost-efficient option for larger farm operators.

Previous to government regulation in 1933 oil extraction decisions were made by companies that leased land from property owners. Typically, oil firms did not own oil fields. Rather, they leased access to fields through a surface point lease. This, however, created common property conditions. Given that a competing firm could access the same field by leasing an adjoining property oil companies had an incentive to maximize their output. The companies therefore frequently raced against each other to drain the same reservoir (Dennis 1985). Consequently, issues of conservation rarely played into production decisions. Limiting one's output simply meant that another company would capture a greater share of a field's oil (Libecap 1989).

As a result, crude oil output and prices were highly volatile. When new fields were discovered oil prices plummeted as production was maximized across the industry. Not long after, prices would climb as known oil fields were pumped dry. This proved both a blessing and a curse for alcohol. On the one hand, it helped create periods of opportunity for alcohol, such as during those "lean" years of oil discovery when crude prices inched upward. Yet it also proved quite debilitating. This was particularly the case during periods when large fields were being brought into production. For reasons just mentioned, such discoveries were typically followed by a spike in oil output and a drop in its market price (after the East Texas field was discovered gasoline prices fell in some parts of the country to as low as two and one-eighth cents a gallon [Harbeson 1940]). In the end, the market never remained optimal for the alcohol industry long enough for it to establish an infrastructure and organizational configuration similar to that previously established by the oil industry. Unfortunately for alcohol, a steady discovery of fields in Kansas, Oklahoma, and Texas throughout the 1920s and 1930s meant that those "lean" years of discovery for the oil industry were few and far between (though California was also a large oil producing state at this time its isolation kept it from supplying the national market until after World War II [Libecap 1989]).

Table 3.1 presents both the nominal and real prices (constant 2005 US dollars) of oil per barrel from 1860 to 1945 (prices in bold indicate a nominal price below one dollar a barrel). The table is insightful on a number of fronts. For one, it adds further context to what has already been discussed. For example, the table illustrates that the price of oil remained high throughout the 1860s and 1870s, spiking right as the Revenue Act of 1862 was passed (which mandated the $2.08 "sin tax" on all alcohol). It is quite possible that alcohol fuel would have been competitive with petroleum fuels during these years had denatured alcohol been exempt from taxation. Indeed, the oil industry spent much of the latter decades of the nineteenth century overcoming extraction, shipping, and refining difficulties (Williamson and Daum 1959). Yet, by the turn of the century it developed efficiencies and scales of economy unmatched by its competitors in other countries (Williamson and Daum 1959). Conversely, had there been a Standard Oil Trust equivalent in the alcohol industry—some centralized organizational force to establish early economies of scale and concentrate lobbying efforts—ethyl alcohol might have been better positioned to capitalize on oil's early production cycles. Note, for example, how inexpensive oil was in 1906 (and in the ensuing years), which marks when industrial alcohol was finally freed from the tax. By the 1930s, oil prices began to stabilize, just when the alcohol movement was finding its organizational feet in the form of the chemurgic movement. This price stabilization was an effect of an interstate oil cartel that formally took shape in 1935.

As oil fields age they become more expensive to operate. The reservoirs in Kansas and Oklahoma in the late 1920s and early 1930s cost considerably more to run than newly discovered fields in Texas (for reasons related to their geology). And with the discovery of the massive East Texas field in 1930 some of these higher cost wells became vulnerable to the price effects of a market now flooded with cheap crude. The oil industry became concerned that "old" wells would have to be closed in response to declining prices, believing greater coordination across firms was needed to ensure its long term economic sustainability (Libecap 1989). This coordination took the form of an interstate oil cartel, which formed with the signing of the Interstate Oil Compact in 1935. As Libecap (1989: 840) explains:

> "The elements [of this cartel] include state prorationing rules to set monthly production totals and to allocate them among regulated wells; Bureau of Mines market-demand estimates for determining state production levels; the Connally Hot Oil Act for federal enforcement of state production rules in interstate commerce; and the Interstate Oil Compact to coordinate state production policies."

This is relevant to our story for it helped marked the end of the oil production cycles that had existed in the oil industry since its inception. The formation of the interstate oil cartel helped remove the aforementioned market fluctuations that created spaces of opportunity for alcohol. And with this, we have yet another factor that contributed to the eventual "closing" of the debate between gasoline and alcohol.

Table 3.1. Nominal and real prices (constant 2005 US dollars) of oil per barrel, 1860 to 1945

Year	Nominal price (real price)	Year	Nominal price (real price)
1860	**$0.49 ($10.81)**	1903	**$0.86 ($18.96)**
1861	$1.05 ($20.84)	1904	**$0.62 ($13.67)**
1862	$3.15 ($50.69)	1905	**$0.73 ($15.10)**
1863	$8.06 ($102.10)	1906	**$0.72 ($15.00)**
1864	$6.59 ($85.30)	1907	**$0.70 ($14.81)**
1865	$3.74 ($50.61)	1908	**$0.61 ($12.97)**
1866	$2.41 ($34.16)	1909	**$0.61 ($12.97)**
1867	$3.63 ($54.03)	1910	**$0.74 ($15.19)**
1868	$3.64 ($54.18)	1911	**$0.95 ($19.09)**
1869	$3.86 ($60.48)	1912	**$0.81 ($16.04)**
1870	$4.34 ($71.78)	1913	**$0.64 ($12.55)**
1871	$3.64 ($60.48)	1914	$1.10 ($20.05)
1872	$1.83 ($32.05)	1915	$1.56 ($24.20)
1873	$1.17 ($21.11)	1916	$1.98 ($26.19)
1874	$1.35 ($24.36)	1917	$2.01 ($23.06)
1875	$2.56 ($47.63)	1918	$3.07 ($30.44)
1876	$2.42 ($45.03)	1919	$1.73 ($19.26)
1877	$1.19 ($24.43)	1920	$1.61 ($19.11)
1878	**$0.86 ($18.29)**	1921	$1.34 ($15.62)
1879	**$0.95 ($19.50)**	1922	$1.43 ($16.60)
1880	**$0.86 ($17.66)**	1923	$1.68 ($19.03)
1881	**$0.78 ($16.01)**	1924	$1.88 ($21.11)
1882	$1.00 ($21.26)	1925	$1.30 ($14.89)
1883	**$0.84 ($18.52)**	1926	$1.17 ($13.55)
1884	**$0.88 ($19.41)**	1927	$1.27 ($14.72)
1885	**$0.71 ($15.66)**	1928	$1.19 ($14.16)
1886	**$0.67 ($14.77)**	1929	**$0.65 ($8.50)**
1887	**$0.88 ($19.41)**	1930	**$0.87 ($12.68)**
1888	**$0.94 ($20.73)**	1931	**$0.67 ($10.30)**
1889	**$0.87 ($19.19)**	1932	$1.00 ($14.86)
1890	**$0.67 ($14.77)**	1933	**$0.97 ($14.05)**
1891	**$0.56 ($12.35)**	1934	$1.09 ($15.63)
1892	**$0.64 ($13.61)**	1935	$1.18 ($16.32)
1893	**$0.84 ($17.86)**	1936	$1.13 ($15.94)
1894	$1.36 ($28.92)	1937	$1.02 ($14.58)
1895	$1.18 ($25.09)	1938	$1.02 ($14.47)
1896	**$0.79 ($16.80)**	1939	$1.14 ($15.40)
1897	**$0.91 ($19.35)**	1940	$1.19 ($14.50)
1898	$1.29 ($27.43)	1941	$1.20 ($13.78)
1899	$1.19 ($25.30)	1942	$1.21 ($13.68)
1900	**$0.96 ($20.41)**	1943	$1.05 ($11.60)
1901	**$0.80 ($17.01)**	1944	$1.12 ($11.41)
1902	**$0.94 ($20.73)**	1945	$1.90 ($16.91)

Data compiled from *Forbes* (http://www.forbes.com/static_html/oil/2004/oil.shtml).

As should now be clear, gasoline's "superiority" over alcohol was *produced*; a characteristic that was birthed from a socio-technical system organizing around gasoline. Take the case of how the auto industry responded to the problem of engine knock. Knocking occurs when combustion of the air/fuel mixture occurs unevenly in the cylinder. The existence of multiple combustion fronts produces a shock wave within the cylinder, resulting in what is commonly known as "engine knock" (Taylor 1985). One of the major problems with early gasoline (pre-tetraethyl lead) was that it unevenly combusted in the cylinder creating disruptions in power, foul smelling exhaust, and this engine knock. Because of its low octane, engineers could not develop a high compression ratio engine to run on conventional gasoline. A resolution to this problem, which had become apparent in the early years of the twentieth century, was blending alcohol with gasoline. Studies consistently revealed that alcohol boosted octane sufficiently to eliminate engine knock (by creating more complete combustion in the cylinder) and therefore produced an overall cleaner running engine (see e.g., Sorel 1907; Tunison 1920).

While automobile engineers experimented early on with different engine configurations (e.g., steam, diesel, and alcohol) (Hard and Jamison 1997), their designs began to stabilize around gasoline during the first decade of the twentieth century. This design closure, however, was not an effect of gasoline being universally recognized as superior to alcohol. Rather, just the opposite was the case. Engines were being built with low compression ratios so as to accommodate the low octane gasoline of the day. The carburetors of these engines thus needed adjustment if they were to optimally operate on alcohol blends (without adjustments their efficiency and power were greatly compromised) (Sorel 1907). For example, the Hart-Part Company of Charles City, Iowa, began equipping some of their tractors with alcohol burning carburetors in 1908 (Wik 1962). That same year Ford began production of his famous Model T, which was designed to run on alcohol, gasoline, or an alcohol blend (Pahl 2005). Olds Gas Power Company, a competing auto manufacturer, soon followed suit, offering an add on component to their cars' carburetor, which would allow them to run on either alcohol or gasoline (Berton et al. 1982). And in 1917, Henry Ford began promoting his new Fordson Tractor, which was designed to run on either alcohol or gasoline (Wik 1962).

Not long after Ford began mass producing the famous Model T, alcohol enthusiasts had hopes that engines would become increasingly designed around this fuel: "It also seems reasonable to expect a greater general improvement in alcohol engines than in gasoline engines" (Clough 1909: 607). In truth, just the opposite happened. In order to handle either fuel, early automobiles (and tractors, motorized boats, etc.) had to be designed to handle the poorer of the two: namely, gasoline. This ultimately worked against alcohol fuel. Having been designed for gasoline, lower compression engines, as one would expect, did not perform as well when run on alcohol.

The term technological momentum is useful here. Developed first by Thomas Hughes (1969), technological momentum reconciles the conceptual tension between technological determinism and social determinism. The former, found in some "strong" applications of the path dependency model (David 2000), claims that the introduction of a technology "locks in" society to a particular future trajectory. Following this model, a technology's existence, after this lock in has occurred, becomes self-sustaining and, by "locking out" alternatives, irreversible (see e.g., Brian 1989). Social determinism, conversely, argues that society alone controls the emergence and future trajectories of technologies, irrespective of the structuring realities of sunken capital, scales of economy, and other material transformations linked to a technology's introduction into society (see e.g., Green 2002).

Technological momentum brings these competing models together by introducing time into the analysis. According to Hughes, society has the greatest control over a technology when it is first introduced. As a technology matures, however, and becomes embedded within society (and society becomes further embedded within it) additional logics begin to play a role in shaping its future course. The metaphor of "momentum" is useful here because it conveys the existence of supra-agent forces—that is, forces that are still social at their core but which cannot be socially constructed away—without reducing the future to these forces.

Closure around gasoline was a function of increasing momentum gathering behind this fuel. Early internal combustion engines being designed to meet the needs of gasoline; the 40 year head-start provided by the alcohol "sin" tax (from 1862 to 1906) that allowed the oil industry to develop a mature production and distribution infrastructure; and the formation of an oil cartel (namely: the Interstate Oil Compact) in 1935 to stabilize the price of crude to eliminate "boom and bust" cycles that provided market openings for alcohol: all represent examples of a socio-technical system locking in around gasoline and therefore locking out alcohol.

There remained a degree of interpretive flexibility as to what constituted a "working" automobile in the early decades of the twentieth century. Initially, the automobile represented function over form. Henry Ford's highly economical Model T provided few frills. Before the 1920s, when Ford reigned supreme among automobile manufactures, the automobile represented a way to get from one place to another. GM, with Alfred Sloan at the helm, challenged Ford's supremacy by reframing "transportation". Their strategy: getting people to believe that getting from one place to another was not enough. Getting there in style, comfort, and with power slowly become part of the country's collective imagination when the thought of transportation was evoked. A "working" automobile was thus no longer a car that simply carried individuals across space; not if it did so while knocking, spewing noxious exhaust, jostling passengers, and lacking sufficient get-up-and-go. Alfred Sloan explained that their strategy called for creating demand "not for basic transportation, but for progress in new cars for comfort, convenience, power and style" (as quoted in Rosner and Markowitz 1985: 344).

As incomes rose people began to accept this new frame and the more expensive cars (and of course Ethyl gasoline) that "fit" this new understanding of transportation. This emphasis on style, power, and efficiency created a new language for automobile manufactures and automobile consumers. A word previously unknown to the automobile world emerged in the 1920s to describe cars (read: the Model T) of the previous decade: old.

Soon alcohol would meet a similar discursive fate. From the distance of almost a century some might read alcohol's disappearance from the fuel market as a case of technocratic optimism landing on the losing side of a struggle for the popular imagination. Remember, until the late 1930s chemurgy was a popular branch of chemistry, often viewed at the cutting edge of the discipline. Thus, unlike early steam cars, which came to represent "old fashionedness" (Hard and Jamison 1997: 148), early biofuels rested upon "the miracles of modern chemistry" (as an article in *Popular Mechanics* [1933: 488] titled "Miracles from Test Tubes" proclaims). For a time, particularly during the 1930s, chemurgy represented the forefront of chemistry in the US (Beeman 1994). But this status was short-lived as chemists turned increasing toward petroleum. "By the late 1930s", as one historical account of plastics in the US explains, "the push to exploit basic chemical raw materials derived from coal, natural gas, or petroleum" had reached a fevered pitch (Meikle 1995: 82).

To look only at the 1930s, one might draw the conclusion that the "state of the art"—namely, chemurgy—lost the discursive struggle over popular imagination. Yet, I believe it more accurate to say that public perceptions of what constituted "state of the art" shifted as landscape conditions changed (e.g., design demands of World War II, advances in synthetic organic chemistry [Steen 2001], etc.). Those on the side of what was perceived to represent "cutting edge" science still, in the end, prevailed.

REFERENCES

Anderson, J. L. 2004. "'The Quickest way Possible': Iowa Farm Families and Tractor-Drawn Combines, 1940-1960," *Agricultural History* 76(4): 669-88.

Beeman, Randal. (1994) 'Chemivisions: The Forgotten Promise of the Chemurgy Movement', *Agricultural History* 68(4): 23-45.

Berton, Hal, William Kovarik and Scott Sklar (1982) *The Forbidden Fuel: Power Alcohol in the Twentieth Century* (New York: Boyd Griffin).

Berk, Gerald. 1994. *Alternative Tracks: The Constitution of American Industrial Order, 1865-1917*. Baltimore, MD: Johns Hopkins University Press.

Bijker, Wiebe 1995 *Of Bicycles, Bakelites, and Bulbs: Toward a Theory of Sociotechnical Change*. Cambridge, MA: MIT Press.

Bolles, Albert 1878. *Industrial history of the United States*. Norwich, CT: Henry Bill Publishing.

Brian, Arthur. 1989. Competing technologies, increasing returns, and lock in by historical events, *Economic Journal* 99: 116-131.

Callon, M. 1995. Technological conception and adoption network: Lessons for the CTA practitioner, In *Managing technology in society: The approach of constructive technology assessment*, edited by A. Rip, T. Misa, and J. Schot, pp. 307-30, New York: Pinter.

Clough, Albert. (1909) 'Comparison of Gasoline and Alcohol Motor Tests', *The Horseless Age* 24(21): 607.

Colbert, T. 2002. "Iowa Farmers and Mechanical Corn Pickers, 1900-1952" *Agricultural History* 74: 530-44.

David, Paul. 2000. Path dependence, its critics and the quest for "historical economics," In Geoffrey Martin Hodgson (eds), pp. 120-44, *The Evolution of Economic Institutions*, Cheltenham, England: Edward Elgar Publishing.

Dennis, Michael. (1985) 'Drilling for Dollars: The Making of US Petroleum Reserve Estimates, 1921-25', *Social Studies of Science* 15(2): 241-65.

Feenberg, Andrew 1999. *Questioning Technology* New York: Routledge.

Gates, David. 1907. Making a Servant of Alcohol: An Account of the Uses and Possibilities of Denatured Alcohol, *The World To-Day* 12(1): 57-62.

Green, Lelia. 2002. *Technoculture: From Alphabet to Cybersex*. Crows Nest, Australia: Allen & Unwin.

Hamlett, P. 1992. *Understanding technological politics*. Englewood Cliffs, NJ: Prentice Hall.

Harbeson, Robert 1940. 'The Present Status of the Sherman Act', *Michigan Law Review* 39(2): 189-212.

Hard M. and A. Jamison (1997) 'Alternative Cars: The Contrasting Stories of Steam and Diesel Automotive Engines', *Technology in Society* 19(2): 145-60.

Hughes, Thomas 1969. Technological Momentum in History: Hydrogenation in Germany 1898-1933, *Past and Present* 44 (1): 106-132.

Kovarik, Bill. 1998. Henry Ford, Charles F. Ketering and the Fuel of the Future, *Automotive History Review* 32: 7-27.

Kovarik, Bill. 2003. "Ethyl: The 1920s Environmental Conflict over Leaded Gasoline and Alternative Fuels," Paper to the American Society for Environmental History, Annual Conference, March 26-30, Providence, RI, http://www.runet.edu/~wkovarik/papers/ethylconflict.html, last accessed December 7, 2007.

Kovarik, Bill. 2005. "Ethyl-Leaded Gasoline: How a Classic Occupational Disease Became an International Public Health Disaster," *International Journal of Environmental Health* 11(4): 384-397.

Kim, Eun-Sung. 2008. Chemical sunset: Technological inflexibility and designing an intelligent precautionary 'polluter pays' principle, *Science, Technology, and Human Values* 33(4): 459-79.

Kramer, Randall 1983. "Federal Crop Insurance1938-82," *Agricultural History* 57(2): 181-200.

Libecap, Gary. (1989) 'The Political Economy of Crude Oil Cartelization in the United States, 1933-1972', *The Journal of Economic History* 49(4): 833-55.

Loeb, Alan 1995. "Birth of the Kettering Doctrine: Fordism, Sloanism and the Discovery of Tetraethyl Lead," *Business and Economic History* 24(1): 72-87,

MacKenzie, Donald 1990. *Inventing Accuracy: A Historical Sociology of Nuclear Missile Guidance*. Cambridge, MA: MIT Press.

Meikle, J. (1995) *American Plastic: A Cultural History* (New Brunswick, NJ: Rutgers University Press).

Pinch, Trevor and Wiebe Bijker 1984. "The Social Construction of Facts and Artifacts: Or How the Sociology of Science and the Sociology of Technology Might Benefit from Each Other," *Social Studies of Science* 14: 399-441.

Popular Mechanics (1933) 'Miracles from Test Tubes', 60 (October): 488-492.

Rasmussen,Wayne. 1962. "The Impact of Technological Change on American Agriculture, 1862-1962," *The Journal of Economic History* 22(4): 578-91.

Robert, Joseph. 1983. *Ethyl: A History of the Corporation and the People WhoMade It* Charlottesville, VA: University Press of Virginia.

Saloutos, Theodore. 1974. "New Deal Agricultural Policy: An Evaluatioan," *The Journal of American History* 61(2): 394-416.

Sargen, Nicholas 1979. *"Tractorization" in the United States and Its Relevance for the Developing Countries* New York: Garland Publishing

Sclove, R. 1995. *Democracy and technology.* New York: Guilford. Taleb, Nassim. 2008. *The Black Swan: The Impact of the Highly Improbable.* New York: Random House.

Shepherd, Geoffrey 1947 *Agricultural Price Policy* Ames, IA: Iowa State College Press.

Sorel, Ernest. (1907) *Carbureting and Combustion in Alcohol Engines* (New York: John Wiley and Sons).

Steen, K. (2001) 'Patents, Patriotism, and 'Skilled in the Art': USA v. The Chemical Foundation, Inc., 1923-26', *Isis* 92(1): 91-122.

Taylor, Charles. (1985) *Internal Combustion Engine in Theory and Practice: Vol 2, Revised Edition* (Cambridge, MA: MIT Press).

Tunison, B. 1920. "The Future of Industrial Alcohol," *The Journal of Industrial and Engineering Chemistry* 12(4): 370-6.

Wik, Reynold (1962) 'Henry Ford's Science and Technology for Rural America', *Technology and Culture* 3(3): 247-258.

Williamson, H. and A. Daum (1959) *The American Petroleum Industry: The Age of Illumination, 1859-1899* (Evanston, IL: Northwestern University Press).

In: A Sociological Look at Biofuels
M. S. Carolan, pp. 45-59

ISBN: 978-1-60876-708-3
© 2010 Nova Science Publishers, Inc.

Chapter 4

ETHANOL'S COMEBACK

Ethanol fell out of favor in the decades following World War II. High corn prices—the price of corn in the 1950s averaged $1.35 per bushel compared to an average price of $2.37 per bushel between 2000 and 2004—made ethanol too expensive to produce. Also, rising petroleum reserves—at least until the 1970s when oil production per capita peaked in the US—kept most from believing a gasoline alternative was necessary. All this changed in the 1970s when corn prices began to drop, the term "peak oil" became a household term, and the price of oil soared (helped in part by the OPEC oil embargo). With the help of government subsidies, ethyl alcohol found its way back into gas stations in the late 1970s. Now called "gasohol" (a gasoline blend containing 10 percent ethanol), corn-based alcohol received its share of fanfare when reintroduced into the Midwest. As announced in the July 27th, 1978 issue of *The Carlisle Citizen* (Carlisle is a small town in Iowa located just Southeast of Des Moines), in an article titled "Five More Iowa Stations Sell Gasohol":

> "Gasohol, using alcohol derived from corn, can help reduce this surplus and hopefully raise corn prices a little. [...] [T]he five stations will probably price Gasohol a couple of cents higher than no-lead regular gasoline [...] but the 8 1/2 cent-per-gallon tax credit for Gasohol will keep the price comparable to no-lead gasoline. A brief kick-off ceremony at 9:45a.m. will precede Gasohol sales at the Spencer Mobil Car Wash. City and state representatives will attend, as well as a representative of the Iowa Corn Growers Association and the Iowa Development Commission" (The Carlisle Citizen 1978: 4).

What happened in the latter decades of the twentieth century, however, cannot be understood as a mere repeat of the debate that occurred between ethyl alcohol and gasoline decades earlier. Unlike in the early decades of the century, gasoline, by the 1970s, was fully entrenched, as evidenced by the degree to which society had come to organize itself around this fuel. In the previous chapter I introduced the concept of the socio-technical system, which serves as a reminder that technologies affect and are an effect of their broader infrastructural, organizational, regulatory, and symbolic environments. Evidence suggests that today's ethanol boom better represents a transition in the automobile socio-technical system than the emergence of anything new. Indeed, as detailed in this chapter, part of ethanol's success can be attributed to how well it fits within existing dimensions of a preexisting system. In order to explain this transition I go back a few decades; a period that, in hindsight, served an important "incubation" function for ethanol by reducing the social and economic transaction costs of producing and consuming this fuel.

The early 1980s saw a retraction of some of the pro-ethanol legislation enacted in the 1970s, thanks in large part to the "free market" ideology of the Reagan administration (Berton, Kovarik and Sklar 1982). Even so, ethanol continued to be produced, spurred on by US supported loans (e.g., in 1980 the Energy Security Act provided insured loans for up to 90 percent of construction costs on ethanol plants [Hahn and Cecot 2008]) and corn subsidies. Early pro-ethanol policies provided the industry an important incubation space that helped shield the fuel from market forces. Non-mainstream technological artifacts often require protection from market signals if they are to ever become a formidable threat to the dominant socio-technical system, what is known as incubation space (Kemp, Schot and Hoogma 1998). This protection is needed because new technologies initially have low price/performance ratio (Geels 2007a). And it is because of this space, shielded from market signals, that ethanol blends could (and can) compete with unleaded gasoline.

Since 1970, under the Clean Air Act's Mobile Source Program, the Environmental Protection Agency (EPA) has had the authority to regulate fuels and fuel additives. Additional regulations were added to the Act in 1990 amendments that mandated oxygenates be added to gasoline to reduce smog in high pollution areas. This reformulated gasoline, or RFG, was typically produced by mixing into gasoline either ethanol or methanol that had been processed into MTBE (Methyl Tertiary Butyl Ether). MTBE was initially the preferred oxygenate. Compared to ethanol, it was cheaper to manufacture, has higher energy content, is not water soluble, can be blended at the refinery and shipped through pipelines, and can be used in warm weather without increasing emissions

(Reitze 2007). A high profile case involving ground water contamination in California, however, quickly revealed MTBE as a major public health hazard. MTBE was not only found to be leaking from ground storage tanks, but water contamination had also been linked to motor vehicle and boat exhaust emission deposits (Williams 2001). By the late 1990s, various states—most notably California and New York—banned MTBE as a gasoline additive. This movement away from MTBE culminated with the Energy Policy Act of 2005, which repealed the Clean Air Act oxygenate requirement. In place of this requirement, the act contains a provision that has been a boon for ethanol; namely, that annual production of gasoline contains at least 7.5 billion gallons of "renewable fuel" by 2012. While the bill uses the term "renewable fuel", ethanol has been the largest benefactor of this provision given its prominence in the US renewable fuel market. The Energy Policy Act of 2005 created a level of market demand for ethanol that had previously never existed. And perhaps even more importantly, the act signaled to potential investors the existence of *future demand*—specifically, of at least 7.5 billion gallons by 2012—which increased investor confidence in this still-fledgling industry.

Research into technological change highlights the importance of expectations (see e.g., Van Lente and Rip, 1998; Brown and Michael, 2003). As one analyst notes:

> "Expectations can act as self-fulfilling prophecies, because they guide social actions in technological change. [...] Expectations are then translated into requirements, indicating directions for R&D activities. This creates a 'protected space' for technology development actors, who receive resources to make the expectations come true." (Geels 2007b: 635, 636)

And there are strong signals pointing to a bright biofuel future in the US that, at least in the short term, are calling out "corn ethanol".

Yet even with all these regulatory openings, ethanol continues to encounter a mismatch with the existing regime, which explains the array of additional policies that continue to shield the fuel from market forces. For example, between 1995 and 2005 the government provided approximately $164.7 billion in agricultural subsidies, of which $51.3 billion went to subsidize the production of corn (Philpott 2007). The General Accounting Office (2000) has estimated that over $19 billion in taxpayer support was given to the ethanol industry between 1980 and 2000. While these subsidies were less than those provided to fossil fuels and nuclear power they exceeded all government subsidies when calculated in per unit energy terms (General Accounting Office 2000).

Another significant development that gave ethanol room to grow is the volumetric ethanol excise tax credit. In 2006, approximately $2.5 billion was distributed to gasoline blenders through the tax credit, which provides a $0.51 per gallon credit for every gallon of ethanol blended with gasoline. The federal tax credit was created in 1978 by the Energy Tax Act. While Congress has adjusted the federal tax credit over the years, the tax has always been extended. The Energy Information Administration predicts that annual production of ethanol will exceed 10 billion gallons in 2010. If the entire amount is blended into gasoline, the federal government could incur almost $5 billion annually in direct costs through the tax credit alone (Hahn and Cecot 2008).

In earlier chapters GM and Standard Oil were singled out for the influence they once commanded in matters related to fuel and the automobile, which ultimately allowed them to shape the socio-technical system within which these artifacts are embedded. Unlike a century ago, ethanol now has its own Standard Oil in its corner: Archer Daniel Midland (ADM). The lobbying efforts of ADM have been pointed to by analysts seeking explanation for why ethanol has received such generous support from the federal government in recent decades. While impossible to say how things today would be different had ADM never existed, its presence has most certainly helped give ethanol momentum.

In 1990, the U.S. ethanol industry had a capacity of 1.11 billion gallons a year, with ADM accounting for 55 percent of its capacity. By June 2006, the industry had more than tripled its capacity. While ADM's portion dropped to 21 percent it continues to dominate the industry because its next largest competitors have a share of the market that is an order of magnitude smaller (see Table 4.1). Much of ADM's political influence comes from an alliance with farmers and through the Renewable Fuels Association (RFA) and the American Coalition for Ethanol (a calculation was done a decade ago that estimated every dollar of ADM's profit costs taxpayers approximately $30 [Bovard 1995]). The RFA is reported to have contributed $772,000 to Republican campaigns and $136,500 to Democratic campaigns in 1991 and 1992 (Reitze 2007). From 1988 to mid-1998, ADM also gave $2 million in soft money to Republicans candidates and $1.1 million to Democrats candidates (Reitze 2007).

Judging by the steady stream of federal support for corn production and ethanol processing in recent decades this appears to be money well spent. Every nominated Republican and Democratic candidate for the US presidency since Jimmy Carter has publically pledged their support for corn-based ethanol (while John McCain opposed corn ethanol subsidies in his unsuccessful run in 2000 he changed his tune in 2008). The corn ethanol lobby has been so effective in part because corn production and ethanol processing plants (see Table 4.1) are largely confined to a handful of less populated states (Sperling and Gordon 2009). Yet even these lightly populated states have two senators. So while interests groups in states like California and New York struggle to have a senator's ear there is far less to distract an Iowa senator from, say, listening carefully to requests from the President of the Iowa Corn Growers Association. The fact that Iowa is the first state to select presidential delegates also forces the ethanol issue to the front-and-center of any candidate's platform. The road to the Whitehouse is paved with pro-ethanol rhetoric.

Table 4.1. Top ethanol producers in the US in 2006

Company	State(s)	Capacity hm3
Archer Daniels Midland	IL, IA, NE, MN, ND	4.1
VeraSun Energy	SD, IA	0.87
Hawkeye Renewables	IA	0.84
Aventine Renewable Energy	IL, NE	0.57
Cargill	NE, IA	0.46
Abengoa Bioenergy	NE, KA, NM	0.42
New Energy Corp.	IN	0.38
Global Ethanol/Midwest Grain	IA, MI	0.36
Total US Capacity		19.0

To protect domestic ethanol production there is also a $0.54 per gallon tariff on imported ethanol (save for a few exemptions) (Rajagopal et al. 2007). One way around this tariff would be for U.S. companies to import sugarcane and then process that cane into ethanol. This would appear to make economic sense, if the USDA's estimates are correct that ethanol could be produced from sugarcane at a lower price than if the source material were corn (U.S. Department of Agriculture 2006). U.S. sugar policy, however, has long restricted imports of foreign sugar (which has caused the U.S. price of sugar to be roughly twice the world market price for decades). Given its implications for domestic biofuels, U.S. sugar policy is now of interest to more than the domestic sugar industry. This no doubt only adds to the lobbying force behind retaining the status quo of this protectionist policy. Except now the policy is protectionist in two respects: it protects the domestic sugar industry *and* the domestic (corn-based) ethanol industry.

These spaces of incubation provided ethanol necessary room to establish new expert systems, ways of thinking, and institutional connections, which have all been shown to contribute to the stabilization of an alternative technical artifact (Rip and Kemp 1998). As the gasoline socio-technical system emerged and gained momentum in the early decades of the twentieth century new forms of expertise were required to support associated technologies and solve problems that might arise through their use. For example, existing institutions like the YMCA (Young Men's Christian Association), in collaboration with automobile clubs and firms, formed technical schools that functioned to support the then-fledgling car network (Fink 1970). Expanding socio-technical systems can eventually lead to the creation of new academic disciplines (Nelson and Winter 1982), which in the case of the (gasoline) automobile involved the emergence of, for example, automobile engineering, automotive machinery, and petroleum geology (Unruh 2000).

Over time expert systems can become self-sustaining (Nelson and Winter 1982). They do this by taking on a structural quality, locking in routines and rules of thumb among like-minded individuals that have been trained to understand problems and solutions in certain ways (Nelson 1995; Unruh 2000). This creates a tendency among institutions/experts to frame issues in ways that reflect their expertise, rather than in ways that threaten to render obsolete the underlying socio-technical system (and by implication their status as an expert) (Christensen 1999). An example of this is how the problem of traffic congestion has consistently been "resolved" by experts in the US by calls for more and wider roads (which inevitably only temporally resolves the problem) rather than public transportation.

Research indicates that as firms increase in size they become adverse to radical innovations as inertia forms behind established R&D streams (Acs and Audretsch 2006). This aversion to truly innovative research is due to a realization that a significant change in trajectory would likely render the firm's inventories, expertise, skills, technologies, and equipment obsolete. Dominant firms thus seek to strengthen their position by re-investing in their core competencies, which creates a self-reinforcing feedback that contributes to the locking in of an existing socio-technical system (which in part explains why biofuels were "locked out" from the system for so long)(Vonortas 1997).

Financial institutions also prefer making loans to companies with collateral and an already established market, which benefits dominant firms and their R&D streams. Within such an environment, alternative R&D streams become the interest of venture capital or government research programs (Unruh 2000). In the case of ethanol, the latter provided much of the support throughout the 1970s, 80s and 90s. Protecting this fuel from market forces helped incubate alternative knowledge systems involving biofuels. This eventually led to the creation of learning economies (Agrote and Epple 1990). Research in biofuels began experiencing increasing returns as expert systems slowly formed and solidified around ethanol. Evidence of this can be found in reductions in costs to process this fuel in the last 20 years. As illustrated in Table 4.2, learning economies in enzyme production and the production of ethanol have resulted in lower overall production costs.

Table 4.2.

Early 1980s		2005		Reduction (%)	
	$2005/m3	% of total cost	$2005/m3	% of total cost	
Energy cost	140	58	70	54	50
Labor cost	55	23	16	12	70
Enzyme cost	40	17	10	8	75
Total cost	240	98	130	74	40-50

Note: m3 = cubic meter of ethanol.
Source: Hettinga et al. 2009: 196.

Before the recent boom, biofuels research was largely limited to a handful of Land Grant Universities (e.g., Iowa State University, Michigan State University, etc.). Today, however, we are beginning to witness the emergence of a self-sustaining expert system directed at biofuels, with dozens of universities becoming involved in this field of research. The formation of new disciplines (e.g., Industrial Biosystems Engineering [Chen 2008]), a refocusing of old ones (e.g., Chemical and Biomolecular Engineering), and the creation of peer reviewed journals (e.g., *Biotechnology for Biofuels*)—all are helping to stabilize and lock in a biofuel-oriented expert system. Organizational feet are beginning to form beneath these renewable fuels, which makes the long-term stability of this expert system all the more likely.

There has also been an incentive in recent years, from the standpoint of university administration, to "bet" on the ethanol juggernaut. Since 9/11, a steady stream of money, from both governmental and non-governmental sources, is being directed at biofuels. Baylor University, for example, was recently awarded a $492,000 grant by the U.S. Department of Agriculture (USDA) to assist in their research on cellulosic ethanol (Biofuels Digest 2008). Other recent USDA grants into biofuels include an $840,000 award to Washington State University for research into phenols in poplar trees (phenols have similar properties to petrochemicals) and a $50 million award to Michigan State University to further their research into ethanol (Biofuels Digest 2008). In 2008, Iowa State University received a $944,000 grant from the US Department of Energy (DOE) to support a project that uses pyrolysis, gasification and nanotechnology based catalyzation to produce ethanol (Biofuels Digest 2008). Corporations have likewise began to fund research into biofuels: British Petroleum, $500 million to a UC-Berkeley lead consortium; Exxon Mobile, $100 million to Stanford University; Chevron, $25 million to UC-Davis; Conoco Phillips, $22.5 million to Iowa State University; Chevron, $12 million to Georgia Institute of Technology; Chevron, $(amount not disclosed) to Texas A & M University (Sheridan 2007). Then there was the $369,000 donation from Wal-Mart to the Arkansas Biosciences Institute at Arkansas State University, which complemented a $1.48 million US DOE grant to support cellulosic ethanol production research at the university (Christiansen 2008).

To be clear, ethanol's success is not displacing conventional automobile expert networks; it is only causing them to shift to meet the demands of a post-breakthrough socio-technical system. The recent biofuel boom does not, for example, require automobile mechanics to learn an entirely new skill set. Ethanol blends do not displace the internal combustion engine, which means this fuel does not threaten, say, the discipline of automotive engineering nor does it diminish the system's need for the automobile machinist or petroleum geologist. Ethanol's breakthrough does, however, alter what some of the "needs" are that these experts are expected to address. For example, engineers must now deal with the needs of fuel containing higher levels of ethanol (E85 blends) when designing engines and fuel distribution systems. Ultimately, while the post-breakthrough automobile socio-technical system contains some new expert roles and network configurations, many others still look the same.

As detailed in earlier chapters, the oil and automobile industries have historically been powerful foes of biofuels. In recent years, however, their animosity towards biofuels has waned. As indicated by the aforementioned industry-university biofuel partnerships, the oil industry has adjusted its research stream to include non-petroleum based fuels. In part, the petroleum industry's shift in attitudes towards biofuels reflects a diversification of R&D, where long term economic sustainability (in a post-peak world) remains the goal. Because of this the oil industry does not view the situation as win-lose. As Rick Zalesky, Vice President of Biofuels and Hydrogen at Chevron Technology Ventures, notes when discussing his company's position towards biofuels: "What helps a lot is that if I sell a gallon of ethanol today, it didn't mean I didn't sell gas. There's growth. It's a bigger pie" (as quoted in Sheridan 2007: 1201). This position is reflected in a recent background report to Congress, which explains that "diluted blends of ethanol, such as E10, are considered to be 'extenders' of gasoline, as opposed to alternatives" (Yacobucci 2007: 8)

The automobile industry too has become noticeably less antagonistic towards ethanol. This warming of relations is no doubt due, at least in part, to how well ethanol "fits" within the existing system and therefore does not represent a threat to the internal combustion automobile. For biofuel proponents and the automobile industry, the solution to the problems of high priced/post-peak oil, global climate change, and energy dependence on nondemocratic states is alternative *fuels* and not alternatives *to* fuel.

Since 1975, the US government has required that automobile manufacturers meet the nation's CAFÉ (Corporate Average Fuel Economy) mileage standards or face penalties. With the passing of the Alternative Motor Fuels Act of 1988, car manufactures receive credit towards meeting the CAFÉ requirement with production of flex-fuel vehicles (FFVs) (automobiles capable of running on both gasoline and alternative fuels). Consequently, from 1993 to 2004, manufacturers were able to increase their CAFÉ by up to 1.2 mpg by producing these FFVs (Szklo, Schaeffer, and Delgado 2007). In other words: each FFV produced helps manufacturers offset the higher mileage autos they are also producing. This offset strategy has become particularly effective when FFVs are used in rental fleets, which has helped inflate production numbers of these vehicles and in turn allowed the manufacture to produce a greater number of less-efficient automobiles (Ries 2006). Importantly, the CAFÉ calculation does not take into account if the FFVs are actually being run on E85, which is fortunate for manufacturers given that of the 170,000 service stations in the United States only 1,200, according to the Environmental Protection Agency, offer E85.[1] Thus, "the main reason why there are over six million flexfuel vehicles registered in the US has less to do with consumer-driven demand than with the long-established policy that credits these vehicles with artificially high fuel economy ratings" (Szklo et al. 2007: 5418).

From a socio-technical standpoint, FFVs provide one solution to the "chicken and egg" problem associated with any shift toward an alternative fuel. These vehicles have been said to offer a bridge between socio-technical systems, pre and post breakthrough (Yacobucci 2007), creating an incubation space to allow manufactures to (quite literally) practice making alternative fuel vehicles. Importantly, however, this practice was limited by the engineering constraint that these vehicles still had to run on conventional gasoline. Thus, while providing a solution to the chicken and egg dilemma, FFVs ensure that both chicken and egg look similar to previous generations of chickens. This provides socio-technical context for the aforementioned quote from the Chevron executive who noted that "if I sell a gallon of ethanol today, it didn't mean I didn't sell gas". While FFVs indicate a degree of flexibility in the socio-technical system they equally show the system's resilience. The expanding presence of FFVs constrain the possibility of viable design solutions, as engineers and designers are forced to deal with the realities of making an engine that operates on *both* gasoline and ethanol. In time, this may lead to the emergence of "rules of thumb", routines, and organization networks that slowly "lock in" design trajectories around duel fuel engines. The bridging function of FFVs, in other words, from pre to post breakthrough, may be more rhetoric than reality.

To put it in very general terms, biofuels are far more profitable today than was the case a century ago. For instance, whereas the issue of patentability made ethyl alcohol less attractive to capital the first time around the same cannot be said today. Monsanto, for instance, has developed genetically modified (GM) sugarcane that is resistant to its Roundup Ready herbicide. This product will be marketed heavily in Brazil with its rapidly expanding sugarcane-biofuel market. Monsanto will also soon begin selling a GM maize variety with high starch content, designed specifically for ethanol production (Suppan 2007).

Another strategy by agribusiness has been to introduce biomass conversion enzymes directly into plants (see, e.g., Sticklen 2006; Torney et al. 2007). Engineering plants to carry the microbial cellulose enzymes that facilitate the conversion of fermentable sugars to ethanol will make these patented plants appreciably more attractive to the biofuel industry than other nonpatented varieties. In fact, one could see a future where *only* enzyme-enhanced plants (read: patented) are allowed to enter the global biofuel stream. In this case, non-engineered varieties, which do not contain conversion enzymes, would suddenly cease to "fit" the system.

Part of ethanol's ascent in the years immediately following September 11, 2001 can be attributed to how it resonated with the public and politicians. Post 9/11 sentiments, calls for energy independence, rising petroleum prices, and global climate change—all have enhanced the symbolic capital attributed to biofuels. As the world's largest producer of ethanol, the US was well positioned to capitalize on recent (largely positive) public sentiments towards renewable fuels. "Problem redefinition" (Bijker 1995: 278) speaks to the importance of framing problems, recognizing that how problems are perceived shape what solutions are ultimately proposed and the socio-technical systems that eventually emerge. As Hughes (1987: 53) explains, "technological systems [emerge to] solve problems or fulfill goals." As detailed in earlier chapters, ethanol proponents have a long history of emphasizing the non-renewable, non-domestic, polluting nature of oil. While this discourse had some traction in the early decades of the twentieth century and again in the 1970s, recent shifts in the global landscape have helped make this discourse transformative. For its part, the oil industry has become less fervent in its denial of these problems, seeing, as they now do, a place for biofuels in the future (Caesar, Riese, and Seitz 2007). This distinguishes the debate today from that of a century ago, when the oil industry was firmly against alcohol fuel. It is not, however, in the oil industry's interest to make the switch anytime soon away from gasoline. Perhaps this is why the oil industry is careful to talk about "renewable" or "bio" fuels and not *alternative* fuels (or alternatives *to* fuel).

The phenomenon of problem redefinition also sets the two eras thus far discussed apart. The following are some of the "problems" that alcohol fuel was looking to address and their eventual "solutions." The problem: engine knock. The solution: adding lead to gasoline. The problem: dwindling oil supply. The solution: massive oil discoveries in Texas in the 1930s. The problem: becoming dependent on other countries for oil (which become particularly acute during World War I when international fuel shipments were disrupted). The solution: massive oil discoveries in Texas in the 1930s. The problem: instability in the agricultural commodity market and depressed grain prices. The solution: New Deal policies that provided a safety net for farmers. As previously discussed, by the late 1930s few of the problems that originally gave alcohol proponents momentum were left to resolve, which offered them little discursive traction upon which to mobilize public support.

Compare this to the "problems" of the first decade of this millennium and how ethanol fits within this definitional scheme. The problem: global warming and greenhouse gases. The problem: dependency upon the Middle East (and unstable countries like Venezuela) for oil. The problem: the overproduction of agricultural commodities and depressed commodity prices (less the case today but a significant issue a few years ago). Against this host of problems, biobased fuels are cast in a notably different light than that which alcohol proponents faced in the first half of the twentieth century.

Yet this positive symbolic capital has recently soured. The "greenness" of ethanol is being challenged as studies report on, for instance, the emissions released during its production, manufacture, transportation, and end-use consumption (Fargione et al. 2008). This could undercut claims regarding ethanol being the "cleaner" fuel that reduces pollution and helps combat global warming. Similarly, to protect the domestic ethanol industry, there is currently a $0.54 per gallon tariff on imported ethanol. If this tax were removed Brazilian ethanol would most certainly begin to flow into the US given the lower production costs of Brazil's sugarcane ethanol (Mathews 2007). But this would also make the US energy dependent on another country and would thus remove another powerful rhetorical devise used by the ethanol industry—namely, that ethanol equals energy independence. Finally, rising global food prices challenge ethanol's status as a renewable fuel (Runge and Senauer 2007). Whereas "renewable fuel" suggests a level of sustainability, rising food prices are causing many to question the long term implications of using this fuel.

It remains to be seen, however, if this change in symbolic meaning will be transformative. There is an inevitable time lag between public opinion and structural change. Investments made in ethanol today are the result of sentiments held yesterday. Similarly, current policies and regulations reflect debates held months and in some cases years ago. While one can take back an opinion about a technological artifact it is much more difficult to take back, say, an ethanol plant, or 1,000 newly manufactured FFVs, or a recently installed E85 pump. Past public sentiments are, quite literally, materializing around us with each passing moment. A disruption of this materialization, while not impossible, will take considerable political will backed by a vocal public.

At the moment—July 2009—gas prices remain far from the record highs experienced last summer (at approximately $2.50 per gallon compared to $4.00 per gallon a year ago). This and a global recession have greatly sidetracked the biofuels debate, at least for the moment. At the same time, decreases in the price of corn (from $8 per bushel in June of 2008 to approximately $3.50 today) and food stuffs more generally (thanks to decreases in transportation costs) have taken some of the edge off the critiques of corn ethanol that a year ago were becoming rather intense.

ENDNOTE

[1] http://www.eere.energy.gov/afdc/fuels/ethanol_locations.html.

REFERENCES

Acs, Zolton and David Audretsch 2006. Innovation in Large and Small Firms: An Empirical Analysis, In *Entrepreneurship, Innovation and Economic Growth*, pp. 3-15, edited by David Audretsch, New York: Edward Elgar.

Agrote, L. and D. Epple 1990. "Learning Curves in Manufacturing," *Science* 247: 920-4.

Berton, Hal, William Kovarik and Scott Sklar 1982 *The Forbidden Fuel: Power Alcohol in the Twentieth Century* (New York: Boyd Griffin).

Bijker, Wiebe (1995) *Of Bicycles, Bakelites, and Bulbs: Toward a Theory of Sociotechnical Change* (Cambridge, MA: MIT Press).

Biofuels Digest. 2008. "USDA awards $492,000 to Baylor to study cellulosic ethanol inhibitors", May 23, < http://www.biofuelsdigest.com/blog2/

2008/05/23/usda-awards-492000-to-baylor-to-study-cellulosic-ethanol-inhibitors/> last accessed September 22, 2008.

Bovard, J. 1995. "A Case Study in Corporate Welfare." *CATO Policy Analysis* 14(241):1-23.

Brown, N., and M. Michael. 2003. The sociology of expectations: Retrospecting prospects and prospecting retrospects. *Technology Analysis & Strategic Management* 15(1): 3-18.

Caesar, W., J. Riese, and T. Seitz 2007. "Betting on Biofuels," *The McKinsey Quarterly* 2: 53-63.

Chen, S. 2008. "Industrial Biosystems Engineering and Biorefinery Systems," *Chinese Journal of Biotechnology* 24(6): 940-5.

Christensen, C., 1999. *Innovation and the General Manager*. Horewood, IL: Irwin

Christiansen, R. 2008. "Wal-Mart donates $369,000 for ethanol research", *Ethanol Producer* October, <http://www.ethanolproducer.com/article-print.jsp?article_id=4777> last accessed September 22, 2008.

Fargione, J., J. Hill, D. Tilman, S. Polasky, and P. Hawthrone 2008. "Land clearing and biofuel carbon debt," *Science* 319 (5867): 1235-7.

Fink, J. 1970. *America Adopts the Automobile Age*. MIT Press: Cambridge, MA.

Geels, Frank 2007a "Analysing the breakthrough of rock 'n' roll (1930–1970): Multi-regime interaction and reconfiguration in the multi-level perspective," *Technological Forecasting and Social Change* 74: 1411-1431.

Geels, Frank 2007b Feelings of discontent and the promise of middle range theory for STS: Examples from technology dynamics, *Science Technology and Human Values* 32(6):

General Accounting Office (GAO). 2000. *Tax Incentives for Petroleum and Ethanol Fuels.* GAO/RCED-00–301R. Retrieved June 29, 2008 (http://www.gao.gov/new.items/rc00301r.pdf)

Hahn, R. and C. Cecot 2008. The Benefits and Costs of Ethanol, AEI Center for Regulatory and Market Studies, Working paper 07-17, Washington, DC.

Hettinga, W., H. Jungingere, S. Dekker, M. Hoogwijk, A. McAloon, K. Hicks. 2009. "Understanding the reductions in US corn ethanol production costs," *Energy Policy* 37: 190-203.

Hughes, Thomas. (1987) 'The Evolution of Large Technological Systems', In W. Bijker, T. Hughes, and T. Pinch (eds.), *The Social Constructions of Technological Systems* (Cambridge, MA: MIT Press): 51-82.

Kemp, R., J. Schot and R. Hoogma 1998. "Regime shifts to sustainability through processes of niche formation: the approach of strategic niche management", *Technological Analysis and Strategic Management* 10: 175–196.

Mathews, J. 2007. "Biofuels: What a Biopact between North and South could achieve," *Energy Policy* 35: 3550-70.

Nelson, R., 1995. "Recent evolutionary theorizing about economic change," *Journal of Economic Literature* 33: 48-90.

Nelson, R. and S. Winter. 1982. *An Evolutionary Theory of Economic Change.* Harvard University Press, Cambridge.

Philpott, Tom. 2007. "The Short Term Solution That Stuck: Where Farm Subsidies Came from and Why They're Still Here." *Grist* January 30. Retrieved December 13, 2007 (http://www.grist.org/comments/ food/ 2007/01/30/farm_bill2/).

Rajagopal, D., S.E. Sexton, D. Roland-Holst, and D. Zilberman. 2007. "Challenge of Biofuel: Filling the Tank without Emptying the Stomach?" *Environmental Research Letters* 2 (Oct.–Dec.). Retrieved March 23, 2008 (http://www.iop.org/EJ/ article/ 1748-h9326/ 2/4/044004/erl7_4_044004.pdf).

Rip, A., and R. Kemp. 1998. Technological change. In *Human choice and Climate Change*, Volume 2, edited by S. Rayner and E.L. Malone, 327-399. Columbus, OH: Battelle Press.

Reitze, Arnold 2007. "Should the Clean Air Act be used to turn Petroleum Addicts into Alcoholics?" *The Environmental Forum* July/August: 50-60.

Ries, R. 2006. "Flex Fuel for Ag?" *Farm Industry News*, September 1, http://www.farmindustrynews.com/mag/farming_flex_fuel_ag/index.html., last accessed March 31, 2008.

Runge, C. and B. Senauer 2007. How biofuels could starve the poor, *Foreign Affairs* 86(3): 41-54.

Sheridan, C. 2007. "Big Oil's Biomass Play," *Nature Biotechnology* 25(11): 1201-3.

Sperling, Daniel and Deborah Gordon 2009. *Two Billion Cars: Driving Towards Sustainability.* Oxford: Oxford University Press.

Sticklen, M. 2006. "Plant Genetic Engineering to Improve Biomass Characteristics for Biofuels." *Current Opinion in Biotechnology* 16:315–19.

Suppan, S. 2007. "Patents: Taken for Granted in Plans for a Global Biofuels Market." Institute for Agriculture and Trade Policy, Minneapolis, MN. Retrieved March 22, 2008 (http://www.iatp.org/iatp/ publications.cfm? refid=100449).

Szklo, A., R. Schaeffer and F. Delgado 2007. "Can one say ethanol is a real threat to gasoline?" *Energy Policy* 35: 5411-21.

The Carlisle Citizen. 1978. Five More Iowa Stations Sell Gasohol. July 27: 4.

Torney, F., L. Moeller, A. Scarpa, and K. Wang. 2007. "Genetic Engineering Approaches to Improved Bioethanol Production from Maize." *Current Opinion in Biotechnology* 18:193–99.

Unruh, G. 2000. Understanding Carbon Lock In, *Energy Policy* 28: 817-30.

U.S. Department of Agriculture (USDA). 2006. The Economic Feasibility of Ethanol Production from Sugar in the United States. Retrieved August 3, 2008 (http://www.usda.gov/oce/EthanolSugarFeasibilityReport3.pdf).

Van Lente, H., and A. Rip. 1998. Expectations in technological developments: An example of prospective structures to be filled in by agency. In *Getting new technologies together*, edited by C. Disco and B.J.R. van der Meulen, pp. 195-220. Berlin and New York:Walter de Gruyter.

Vonortas, Nicholas. 1997. *Cooperative Research and Development*. The Netherlands: Kluwer Academic.

Williams, P. 2001. MTBE in California Drinking Water: An Analysis of Patterns and Trends, *Environmental Forensics* 2(1): 75-85.

Yacobucci, Brent. 2007. "CRS Report for Congress. Fuel Ethanol: Background and Public Policy Issues," Congressional Research Service, Washington, DC, January 24, http://fpc.state.gov/documents/organization/62837.pdf, last accessed March 4, 2008.

In: A Sociological Look at Biofuels
M. S. Carolan, pp. 60-76

ISBN: 978-1-60876-708-3
© 2010 Nova Science Publishers, Inc.

Chapter 5

FROM FOOD TO FEED

I have been careful not to pronounce judgment on corn-based ethanol. With the help of concepts from the subfield of the sociology of technology earlier chapters sought to provide a descriptive and explanatory account of ethanol in the US over the last century and a half. With this chapter I turn a corner by both evaluating biofuels—recognizing that all biofuels are not the same—and offering a speculative look ahead.

To begin I attempt to clear through the ideological and scientific clutter on the subject by answering the question, "What are the costs and benefits of biofuels?". This is an immensely complicated question to answer. Though discussion up to this point has focused on corn-based ethanol, biofuels come in many different forms. To evaluate the costs and benefits of corn-based ethanol we need to understand the alternatives and their costs and benefits: not an easy task. A thorough analysis of *all* biofuels would be cost prohibitive. To achieve a similar end, but do so on the cheap, I review what the scientific peer-reviewed literature says about the costs and benefits of biofuels. A search was conducted on four peer-reviewed journals—*Science*, *Natural Resources Research*, *Renewable Energy*, and *Journal of Clean Production*—from 2006 through 2008 for original research looking into the virtues and drawbacks of biobased fuels. I then assessed this research collectively.

I intentionally avoided conducting a review of the last ten or twenty years of peer-reviewed literature on the costs and benefits of biofuels. Such a review would not give an accurate statement about biofuels *today*. The science involved is evolving rapidly as are the infrastructural capabilities and expert systems, which means analyses into the costs and benefits of biofuels decades ago could easily be outdated today. By restricting my review to the latest literature I hope to arrive at an understanding that is based upon the best available science and the socio-technical realities of the present. The following review includes only original research published in the journals *Science, Natural Resources Research, Renewable Energy*, and *Journal of Clean Production*. The search was limited to research articles published from 2006 through 2008 that evaluated some aspect of this fuel. If the article did not include original research (e.g., Comments from the Editor), it was not analyzed.

Two key terms were searched: "biofuels" and "ethanol" (I learned a search of only "biofuels" often missed articles on ethanol). In the journals *Science* and *Natural Resources Research* a title search was conducted using these terms, while in *Renewable Energy* and *Journal of Clean Production* a title, abstract, and keyword search was used. (The divergence in parameters is an artifact of the search properties offered by each journal.) These search criteria produced 10 articles, which were then analyzed. A more specific breakdown of the search results for each journal is provided in Table 1.

Table 5.1.

Journal	"Biofuel" returns (articles analyzed)	"Ethanol" returns (articles analyzed)
Science	30	5
	(3)	(0)
Natural Resources	0	3
Research	(0)	(2)
Renewable	9	21
Energy	(0)	(1)
Journal of Cleaner	6	10
Production	(2)	(2)

Table 5.2 links up specific articles with the code/s they were assigned during the analysis. Although ten articles were examined, half studied more than one biofuel form. To separate these results I therefore coded for each finding (rather than for each paper). Fifteen "findings" are thus represented in the ten papers analyzed. Table 5.2 also highlights the biomass sources/biofuel types analyzed and the respective country origins of each study.

Articles were analyzed and coded for specific costs and benefits linked to biofuels. Fourteen such dimensions were recorded after reviewing the ten articles. A summary of the results are presented in Table 5.3. As the table notes, articles were also categorized according to whether they examined biofuels derived from foodstock or waste/non-foodstock (one article did not specify the biomass source of the gasoline-alcohol blend examined). To be clear: "foodstock" refers to biofuels developed using products that otherwise would have entered into the food system (either as food or feed); "waste/non-foodstock" refers to fuels derived from products (typically) not part of the food system (e.g., biomass waste, perennials, etc.). Finally, to summarize the findings Table 5.3 presents the "averages" for each dimension for foodstock and non-foodstock derived biofuels. I will now speak briefly about some of the specific findings as they apply to each of the fourteen dimensions.

Table 5.2.

Reference	Biomass Source(s)/Type(s)	Country/ies	ID #s
Fargione et al. 2008	foodstock/non-foodstock	Brazil, Southeast Asia, US	1/14
Patzek 2007	corn ethanol	US	2
Pimentel and Patzek 2007	corn ethanol/sugarcane ethanol	US/Brazil	3/4
Reijnders and Huijbregts 2008	palm oil	South Asia	5
Leng et al. 2008	cassava E10	China	6/7
Nguyen and Gheewala 2008	cassava E10	Thailand	8
Searchinger et al. 2008	corn ethanol/switchgrass	US	9/12
Beer and Grant 2007	wheat/molasses	Australia	10/11
Tilman et al. 2006	perennials	US	13
Topgul et al. 2006	gasoline blends (source unclear)	Turkey	15

BIODIVERSITY

Converting rainforests, peatlands, savannas, or grassland to produce corn, sugarcane, soybeans, and palms for biofuel production has a negative affect on biodiversity (in terms of, for instance, plant, animal, and insect species richness). Conversely, biodiversity was found to be positively affected by the production of waste/non-foodstock fuels, such as when low-input high-diversity biomass (e.g., perennials) is grown on degraded and/or abandoned land. This highlights how policies to significantly replace gasoline with biofuels must be mindful of the issue of biodiversity. If such initiatives translate into monocropping—especially if monocropping is replacing species rich spaces like grasslands or rainforests—then diversity will be negatively affected. It is questionable if a global shift to biofuels can occur without significant land-use change and the loss of biodiversity that would likely follow.

Table 5.3.

Article ID #	1	2	3	4	5	6	7	8	9	10	11	Average (1-11)	12	13	14	Average (12-14)	15
Biodiversity	↓	X	X	X	X	X	X	X	X	X	X	↓	X	↑	↑	↑	X
Global warming	X	X	↑	↑	X	↑	↓	X	↑	X	X	↑	↑	↓	↓	↓	X
CO_2 (net)	↑	↑	↑	↑	↑	X	↑	↓	↑	X	X	↑	↑	↓	NC	mixed	X
CO_2 (auto)	X	X	X	X	X	↓	X	X	X	NC	↓		X	X	X	X	↑
SOX	X	X	X	X	X	↓	↑	X	X	X	X	mixed	X	X	X	X	X
NOX	X	X	X	X	↑	NC	↑	↓	X	X	X	↑	X	X	X	X	X
CO	X	X	X	X	X	X	↑	↓	X	X	X	↑	X	X	X	X	X
Ecological Toxicity	X	↑	↑	↑	X	X	X	X	X	X	X	↑	X	↑	X	↑	X
Hydrocarbons	X	X	X	X	X	X	X	X	↑	↓	↑	↑	X	X	X	X	↑
Water Security	X	X	↑	↑	X	X	X	X	X	X	X	↑	X	↓	X	↓	X
Net energy	X	X	↑	↑	X	X	X	↑	X	X	X	↑	X	↑	X	↑	X
Food displacement	↑	X	↑	↑	X	X	X	X	X	X	X	↑	X	NC	NC	NC	X
Soil erosion	X	X	↑	↑	X	X	X	X	X	X	X	↑	X	↑	X	↑	X
VOCs	X	X	X	X	X	↑	↓	X	X	X	X	mixed	X	X	X	X	X

↑ -Increase impact for biofuel
↓ -Decrease impact for biofuel
NC -No significant change reported
X -Not assessed

GLOBAL WARMING

Biofuels are often labeled by proponents as carbon negative because of net ecosystem carbon dioxide sequestration (Zacharias 2005). When biofuels are produced on recently "opened" land, however, any gains of sequestration are outweighed by greenhouses gasses released when soil is prepared for its first crop. There was not a consensus on the topic among articles reviewed. One article (ID# 6) concluded that, in terms of vehicle emissions, E10 released fewer greenhouse gasses into the atmosphere than if conventional gasoline were used. In terms of *life cycle analysis*, however, the majority seem to believe that biofuels (as currently produced) exacerbate global warming.

CO_2 (NET)

There was significant agreement that, *over its entire lifecycle*, biofuel use leads to an increase in CO_2 emissions—what one study called the "biofuel carbon debt" (Fargione et al. 2008). These studies were concerned about carbon emissions that occur as farmers respond to higher grain prices and convert forest and grasslands into cropland. Interestingly, one article (ID# 12) reports that, in addition to the doubling of CO_2 emissions that result from corn ethanol, switchgrass ethanol, when grown in the US, increases CO_2 emissions by 50 percent. These findings emphasize the value of biofuels from a biomass source that either minimizes land-use change (e.g., waste products) or that can be grown on carbon-poor lands that will not emit massive amounts of CO_2 with land-use change.

CO_2 (AUTO)

There is consensus that biofuels (e.g., ethanol blends) do not increase an automobile's CO_2 emissions. Three of the four studies that examined this characteristic found that ethanol blends reduce the level of CO_2 in a car's exhaust (the other finding noted no change). However, in light of the preceding paragraph, this finding masks significant life cycle CO_2 emissions linked to current biofuels (which is perhaps why biofuel proponents often focus on end-of-tailpipe rather than life cycle emissions [see e.g., Zacharias 2005]). This finding is nevertheless encouraging. If biofuels can be widely produced using sources that have no impact on land-use change then such end-of-tailpipe reductions become environmentally significant.

SOX

Findings #6 and #7 are derived from the same article. This paper conducted an end-of-tailpipe and life cycle analysis on cassava-based E10 in China. The end-of-tailpipe analysis found a 46 percent reduction in SOX emissions, whereas the life cycle analysis reports an 18 percent increase in SOX emissions (further evidence for why end-of-tailpipe analyses tell only part of the story).

NOX

The studies were mixed in regard to NOX emissions. Two life cycle analyses (ID#s 5 and 7) reported an increase in NOX emissions, one (ID# 8) reported a decrease in NOX emissions, and an end-of-tailpipe analysis (ID# 6) reported no change. It is worth noting that a lack of consensus on the subject of NOX emissions over the course of a biofuel's life cycle does not necessarily indicate scientific disagreement. In this case, the one study (ID# 8) that reports a decrease in NOX emissions involved cassava-based ethanol produced in Thailand, whereas findings #5 and #7 involved South Asia palm oil biodiesel and cassava-based ethanol produced in China respectively. The discrepancy in findings may reflect a divergence in how these fuels are produced, which could account for different NOX emissions.

CO

Only one study examined CO emissions (ID# 8). This article performed a life cycle analysis of E10 fuel produced using cassava-based ethanol. A 15.4 percent decrease of CO was reported (compared to conventional gasoline).

ECOLOGICAL TOXICITY

The heavy historical reliance of fertilizers, pesticides and herbicides to produce soybeans and corn are widely known (Belasco 2006). Sugarcane produced in Brazil is also heavily dependent upon petrochemicals (Pimentel and Patzek 2007). The widespread use of chemical inputs to produce today's biofuels was noted among a handful of reviewed articles as having negative affects on the environment and species biodiversity specifically.

HYDROCARBONS

Two studies (in the same article) examined end-of-tailpipe emissions of hydrocarbons. In one study, E10 fuel (ID# 10) was examined, while the other experiment centered on E85 fuel (ID# 11). Both studies reported an increase in end-of-tailpipe emission of hydrocarbons.

WATER SECURITY

The aforementioned reliance upon petrochemicals was also linked to the pollution of waterways. This threatens the viability of communities—human and non-human—that are dependent upon these water sources. Today's biofuels also require a lot of water. As one study notes, "[b]ecause it takes about 12 kg of sugarcane to produce 1 L of ethanol, thus about 7000 L of water are needed to produce the required 12 kg of sugarcane per liter of ethanol" (Pimentel and Patzek 2007: 240). Conversely, biomass sources that are climatologically suited for the growing region and/or that do not require petrochemical inputs were noted as having no negative affect on water security. Moreover, biofuels that do not result in land-use change—and thus do not lead to the draining of marshlands and plowing of grasslands—would also have little impact on watersheds.

NET ENERGY

There was consensus among reviewed articles on this subject, all noting a net energy gain when biofuels are produced. Findings #3 and #4 were contained in the same article. In this piece, ethanol production from US and Brazilian sugarcane was examined. The authors' conclusion: "[B]ased on all the fossil energy inputs in US sugarcane conversion process, a total of 1.12 kcal of ethanol is produced per 1 kcal of fossil energy expended. In Brazil a total of 1.38 kcal of ethanol is produced per 1kcal of fossil energy expended" (Pimentel and Patzek 2007). Finding # 8 drew upon research examining cassava-based ethanol in Thailand. Finally, Finding # 13 examined low-input high-diversity mixtures of native grassland perennials. Yet, as previous research on the subject has made clear: the biomass source matters in determining net energy gains (or losses). Some forms of biomass are questionable net energy sources, such as US corn-ethanol (Pimentel 2003). Other biomass sources, like molasses-based ethanol in India (Blottnitz and Curran 2007), provide clear net energy gains. When discussing issues related to net energy return, it is important to be clear that not all biofuels are equal.

FOOD DISPLACEMENT

Grains (this includes corn) make up approximately 80 percent of the world's food supply (Pimentel and Pimentel 1996). As more of this foodstock is used as fuelstock global food security becomes increasingly threatened. The negative impact that today's biofuels are having on food supply—such as in terms of creating an upward pressure on food prices—has been widely reported (see e.g., Boddiger 2007; Runge and Senauer 2007). For example, in 2002 the National Center for Policy Analysis calculated that US ethanol production is adding more than $1 billion dollars per year to the cost of beef production. The articles reviewed here that address issues of food displacement/security echoed these concerns. Against this backdrop some articles looked at the impact of non-foodstock biofuels on the food supply. If properly regulated—so as to create incentives to use only field waste or to encourage producers, processors and investors to raise low-input high-diversity perennials on marginal land not otherwise used to produce grains—the food supply could be minimally impacted by biofuels. This finding also provides the following sober reminder: biofuels alone are not the panacea to our addition to oil if we wish for future generations to eat.

SOIL EROSION

Another point of consensus: foodstock biofuels increase soil erosion; non-foodstock sources, such as switchgrass and perennials, reduce soil erosion (if grown on previously cultivated land). As explained in one article:

> "Sugarcane production causes more intense soil erosion than any crop produced in Brazil because the total sugarcane biomass is harvested and processed in ethanol production. This removal of most of the biomass leaves the soil unprotected and exposed to erosion from rainfall and wind energy. For example, soil erosion with sugarcane cultivation is reported to have the highest soil erosion rate in all Brazilian agriculture, averaging 31t/ha/year" (Pimentel and Patzek 2007: 239).

Though not addressed in any of the articles reviewed a cautionary word needs to be made about one non-foodstock: field waste. Like sugarcane, the removal of field waste (e.g., corn stalks), which traditionally remains on the field through winter and is tilled under in the spring, would leave the soil unprotected and exposed to erosion. Table 3 presents a generally positive view of non-foodstock biofuels. Yet, like foodstock biofuels, non-foodstock sources cannot be treated equally. Each has its own virtues and drawbacks.

VOCs

While the results are described as "mixed" on Table 3 this deserves some clarification. Findings #6 and #7 come from the same article, where an end-of-tailpipe (#6) and life cycle analysis (#7) are conduced on cassava-based E10 in China. Thus, Finding #6, where a 12 percent increase was reported, speaks to vehicle operation emissions between E10 and gasoline. Finding #7, on the other hand, where a 13 percent decrease was recorded, references life cycle emissions between E10 and gasoline.

AN ASIDE: DIMENSIONS OFTEN MISSED IN "SCIENTIFIC" TECHNICAL REPORTS

Before concluding this section it is important to remind the reader how scientific analyses both illuminate and mask reality (Carolan 2006). Granted, no analysis can cover everything. We must be careful, however, so as not to make "the problem" an artifact of our scientific measuring stick. Take the scientific analyses reviewed above. The majority look to better understand material costs and benefits of biofuels, in terms of, say, emissions, pollution, soil erosion, and so forth. Rarely, however, were attempts made to analyze the *social* costs of biofuels (though social variables were occasionally mentioned [e.g., Pimentel and Patzek 2007]). This scientific blindspot risks focusing public debate around variables measured while drawing attention away from those ignored (even though these costs are no less real).

For example, a recent United Nations report calculated that as many as 60 million indigenous people worldwide will be driven from their lands as a result of biofuel production (Tauli-Corpuz and Tamang 2007). In Brazil, as many of half a million people work in the tropical sun to harvest sugar cane (Smith and Caminda 2007). And while a major employer, the human rights violations associated with this business are adding up (Gangadharan and Larcadas 2007). In order to meet output targets and compete with mechanized harvesting technologies, the requirements for manual cutters have increased from eight tons of sugarcane per day in the 1980s to twelve tons per day in 2006. In terms of its toll on the human body: for every ten tons of sugarcane cut a worker makes some 72,000 machete blows while having to flex their legs 36,000 times (Raynes 2007). From 2002 to 2005, according to Brazil's Social Security Administration, 312 sugar and ethanol workers died on the job, and 82,995 suffered accidents while working in cane fields and ethanol plants (Smith and Caminda 2007). Labor prosecutors in Brazil are likewise investigating the cases of 21 people who died since 2004 while cutting cane. Most were between the ages of 25 and 35 years old (Smith and Caminda 2007).

I mentioned these social variables so as not to further reify the picture of reality presented in the reviewed articles. Scientists "see" the world through their disciplinary lens (Carolan 2008). We must be mindful, therefore, of the limited view presented in the reviewed technical reports (and in Table 5.3), recognizing that there are still other measurable costs and benefits that need to be part of the public discourse surrounding biofuels.

This review highlights why we cannot treat biofuels monolithically; each has its own virtues and drawbacks (with some clearly having more of one than the other). If any generalized comment can be made about the above findings it is this: non-foodstock biofuels appear to have fewer costs—and more benefits—than the conventional biofuels of today. And it appears that most biofuel proponents would agree with me on this, given their repeated pointing to the forthcoming second generation when the costs associated with current (first generation) biofuels are highlighted (see e.g., Mousdale 2008). This begs the question, which takes us beyond the reviewed articles and into terrain of earlier chapters: what can be said of society's chances of eventually switching over to these alternative biomass sources?

The logic of technological momentum weighs heavily in guiding future biofuel trajectories. When applied to analyzing technologies, this momentum speaks to the "locking in"—and correspondingly to the "locking out" of alternatives—of technological trajectories. The above review should make anyone question the long-term viability of first generation biofuels. The logic of technological momentum, however, will likely slow society's ability to switch away from these problematic fuels.

There is tremendous organizational momentum building behind first generation biofuels, as detailed in the preceding chapter. While discussion has thus far focused on the US, which currently leads the world in ethanol production with some 168 ethanol distilleries that produced more than 9.2 million gallons of ethanol in 2008 (up from 6.5 billion in 2007),[1] similar tendencies driving 1st generation fuel exist throughout the world. Take Brazil, which is experiencing an explosion in investments in ethanol derived from sugarcane. In 2005, the figure reached approximately US$6 billion, which included new plants, acquisitions, and expansions. In 2006, biofuel investments hit US$10 billion. The Brazilian oil company Petrobras announced in the spring of 2009 that it will invest US$2.8 billion in biofuel production including ethanol from sugar cane. This brings an estimated total investment of US$174.4 billion for Brazilian sugar cane ethanol for the period 2009-2013 (Costa 2009). Tens of millions of dollars are likewise being invested annually to develop biofuel industries in such countries as China, India, Indonesian, Malaysia, and Thailand (see, e.g., Biospectrum 2008; U.S. Department of Agriculture 2006). These large amounts of capital sunk annually into particular biofuel infrastructural trajectories—whether it is corn ethanol in the United States, sugarcane ethanol in Brazil and China, or palm-oil biodiesel in Indonesia and Germany—are locking countries into specific biofuel regimes (and locking them out of others). This is not to suggest that these countries cannot go down a different biofuel path. What it does emphasize, however, is the fact that over time such alternative paths will come only at ever-increasing costs, particularly if they involve novel biofuels not well suited for the social, institutional, and infrastructural configurations already in place.

With this amount of capital sunk annually into particular biofuel infrastructural trajectories, we must be mindful of the underlying locking in properties that are steadily making alternatives less attractive. A recent study, for example, notes how industrial production costs of ethanol have declined by 45 percent since 1983 (Hettinga et al. 2009). Economies of scale, learning economies, adaptive expectations, and network economies emerge over time giving technology momentum, which explains the cost reductions in US ethanol production in recent years. Again, unless second generation biofuels can utilize the organizational and infrastructural capabilities created for first generation fuels any switch will only come at great expense.

Then there is the issue of what is called policy path dependency: the well-documented phenomena whereby once a policy is enacted it tends to become "locked in" (Kay 2003). More specific to the topic at hand, agricultural policies (e.g., subsidies) have been shown to accumulate a type of momentum if allowed to persist over time (see e.g., Harvey 2004, Law et al. 2008). A number of factors lie behind this. For instance, the greater the levels and history of support, the more dependent the farm and farm supply chain becomes to continuing levels of support, which creates tremendous resistance to the removal of these policies (Harvey 2004). Agricultural producers having more to lose from the removal of subsidies than any other group would gain suggests that the steps taken to remove government support of current agricultural policies would come only at great political expense. As one economist explains:

> "It follows that producers' individual gains from market protection will outweigh individual consumer and taxpayer losses, incurred as a result of the protection. It will thus pay producers to exert more effort in persuading the political system of their just deserts than consumers and taxpayers can be expected to spend on opposing such protection" (Harvey 2004: 266).

Policy path dependency is worth keeping in mind as talk of second generation biofuels grow. Agricultural subsidies spanning generations have helped prop up today's global biofuel regime, particularly in the US and Brazil (the two largest producers of this fuel). Tremendous resistance will be encountered if these support programs are perceived to be threatened in the hopes of encouraging a shift toward alternative biomass sources.

A further locking in tendency comes from an emerging global socio-technical regime that is directed at first generation biofuels. For centuries, alcohol fuel production and consumption was driven exclusively by local logics (e.g., local needs, available agricultural commodities, etc.). Recently, however, these local logics are being slowly replaced by global drivers. We are seeing the emergence of international biofuel "alliances", such as between the US and Brazil, which encourage a sharing of ethanol related technological know-how between countries (White House 2007). Biofuel trade agreements, such as between Ecuador and the

European Union, are also beginning to take root to improve the global flow of these fuels (Seelke and Yacobucci 2007). Finally, "biofuel pacts" between OECD countries and non-OECD countries are being proposed that will institutionalize and harmonize the global trade of biofuels (Mathews 2007). Taken together, these changing global landscape conditions are helping to form a stabilizing (global) logic around certain biofuel forms.

To be sure, paths can shift and take new directions. And the current landscape conditions are such that it is impossible to predict with any accuracy the future trajectory of the socio-technical system in question. While momentum is clearly building behind first generation biofuels this does not mean a shift to second generation fuels is impossible; only, at least in the short term, improbable. If such a shift were to occur, it would only take place at great expense. The Obama Administration has already put forth a number of policies linking energy more closely with environmental sustainability and energy independence. In addition, new economic and market conditions, such as the near collapse of the US automobile industry, has created an opportunity for massive investments in new—and in some cases "green"—technologies in an attempt to stimulate the economy. For example, if such investments were partially directed at battery technology, which has improved considerably in recent years, all-electric cars could soon become both technologically and economically feasible. The federal government is injecting large amounts of money into the field of bioenergy. Overcoming their SUV hangover, which pushed both Chrysler and GM into bankruptcy in 2009, car manufactures are also renewing their efforts to develop "greener" automobiles. In other words, new incubation spaces are emerging for technologies directed at not only *alternative* fuels but also alternatives *to* fuel.

ENDNOTES

1

http://renewablefuelsassociation.cmail1.com/T/ViewEmail/y/698C04744910BF06

REFERENCES

Beer, T. and T. Grant. 2007. Life cycle analysis of emission from fuel ethanol and blends in Australian heavy and light vehicles, *Journal of Cleaner Production* 15: 833-7.

Belasco, W. 2006. *Meals to Come: A History of the Future of Food*. Berkeley, CA: University of California Press.

Biospectrum 2008. "India, China Move up in E&Y Biofuels Investment Sweepstakes," *Biospectrum: Asia Edition* March 28, http://www.biospectrumasia.com/content/280308IND5877.asp, last accessed April 1, 2008.

Blottnitz, H. and M. Curran 2007. A review of assessments conducted on bio-ethanol as a transportation fuel from a net energy, greenhouse gas, and environmental life cycle perspective, *Journal of Cleaner Production* 15: 607-19.

Boddiger, D. 2007. Boosting biofuel crops could threaten food security, *The Lancet* 370(9591): 923-4.

Carolan. M. 2006. Scientific knowledge and environmental policy: why science needs values, *Environmental Sciences* 3(4): 229-37.

Carolan, M. 2008. The Bright- and Blind-Spots of Science: Why Objective Knowledge is not Enough to Resolve Environmental Controversies, *Critical Sociology* 34(5): 725-40.

Costa, Al. 2009. Biofuels Digests Special Report from the 2009 Ethanol Summit, *Biofuels Digest* June 4, http://biofuelsdigest.com/blog2/2009/06/04/biofuels-digest-special-report-from-j the-2009-ethanol-summit/ last accessed July 6, 2009

Fargione, J., J. Hill, D. Tilman, S. Polasky, and P. Hawthorne. 2008. "Land Clearing and the Biofuel Carbon Debt," *Science* 319: 1235-8.

Gangadharan, Anna and Albert Larcadas 2007. "US Lifting of Tariff on Brazil Ethanol Might Spell Trouble for Amazon and Sugarcane Cutters," *Brazzil Magazine* August 27, http://www.brazzil.com/index2.php?option=com_content&do_pdf=1&id=9960, last accessed March 24, 2008.

Harvey, D. 2004. "Policy dependency and reform: Economic gains versus political pains," *Agricultural Economics* 31: 265-75.

Hettinga, W., H. Junginger, S. Dekker, M. Hoogwijk, A. McAloon, and K. Hicks. 2009. "Understanding the reductions in US corn ethanol production costs: An experience curve approach," *Energy Policy* 37: 190-203.

Kay, A. 2003. "Path dependency and the CAP," *Journal of European Public Policy* 10(3): 405-20.

Leng, R., C. Wang, C. Zhang, D. Dai, G. Pu. 2008. Life cycle inventory and energy analysis of cassava-based fuel ethanol in China, *Journal of Cleaner Production* 16: 374-84.

Law, M., J. Tonon, and G. Miller. 2008. "Earmarked: The political economy of agricultural research appropriations," *Review of Agricultural Economics* 30(2): 194-213.

Mathews, J. 2007. "Biofuels: What a Biopact between North and South could achieve," *Energy Policy* 35: 3550-70.

Mousdale, D. 2008. *Biofuels: Biotechnology, Chemistry, and Sustainable Development*. Boca Raton, FL: CRC Press.

National Center for Policy Analysis. 2002. Ethanol Subsidies. Idea House. National Center for Policy Analysis. http://www.ncpa.org/pd/ag/ag6.html, last accessed March 30, 2008.

Nguyen, T. and S. Gheewala. 2008. Fossil energy, environmental and cost performance of ethanol in Thailand, *Journal of Cleaner Production* 16: 1814-21.

Patzek, T. 2007. "A First-Law Thermodynamic Analysis of the Corn-Ethanol Cycle," *Natural Resources Research* 15(4): 255-70.

Pimentel, D. 2003. Ethanol Fuels: Energy Balance, Economics, and Environmental Impacts Are Negative, *Natural Resources Research* 12(2): 127-34.

Pimentel, D. and T. Patzek. 2007. Ethanol production: energy and economic issues related to US and Brazilian Sugarcane, *Natural Resources Research* 16(3): 235-42.

Pimentel, D. and M. Pimentel 1996. *Food, Energy and Society* Boulder, CO: Colorado University Press.

Raynes, Emma. 2007. "The Larger Story: Sugar Cane Workers and Ethanol," Center for Documentary Studies, Duke University, Durham, NC, http://cds.aas.duke.edu/hine/raynes.html, last accessed March 11, 2008.

Reijnders, L. and M. Huijbregts. 2008. Palm oil and the emission of carbon-based greenhouse gases, *Journal of Cleaner Energy* 16: 377-832.

Runge, C. and Benjamin Senauer. 2007. "How Biofuels Could Starve the Poor," *Foreign Affairs* 86(3), http://www.foreignaffairs.org/20070501faessay86305/c-ford-runge-benjamin-senauer/how-biofuels-could-starve-the-poor.html, last accessed December 12, 2007.

Searchinger, T., R. Heimlich, R. Houghton, F. Dong, A. Elobeid, J. Fabiosa, S. Tokgoz, D. Hayes, T. Yu. 2008. Use of US croplands for biofuels increases greenhouse gases through emissions from land-use change, *Science* 319: 1238-40.

Seelke, C. and B. Yacobucci. 2007. "Ethanol and Other Biofuels: Potential for US-Brazil Energy Coorperation." Congressional Research Service, Washington, DC, September 27. Retrieved March 31, 2008 (http://www.wilsoncenter.org/news/docs/CRS%20Report%20on%20US-hBrazil%20potential%20cooperation%20on%20biofuels.pdf).

Smith, Michael and Carols Caminda. 2007. "Ethanol's Deadly Brew," *Bloomberg.com* October 4, http://www.bloomberg.com/news/marketsmag/mm_1107_story3.html#, last accessed March 24, 2008.

Tauli-Corpuz, Victoria and Parshuram Tamang . 2007. "Oil Palm and Other Commercial Tree Plantations, Monocropping and the Impacts on Indigenous Peoples' Land Tenure and Resource Management Systems and Livelihoods," Special Rapporteur, United Nations Permanent Forum on Indigenous Issues, May, http://www.un.org/esa/socdev/unpfii/documents/6session_crp6.doc, last accessed March 24, 2008.

Tilman, D., J. Hill, and C. Lehman. 2006. Carbon-negative biofuels from low-input high-diversity grassland biomass, *Science* 314: 1598-1600.

Topgul, T., H. Yucesu, C. Cinar, and A. Koca. 2006. The effects of ethanol-unleaded gasoline blends and ignition timing on engine performance and exhaust emissions, *Renewable Energy* 31: 2534-42.

United States Department of Agriculture 2006. "China, Peoples Republic of Bio-Fuels: An Alternative Future for Agriculture 2006," United States Department of Agriculture, Foreign Agricultural Service, GAIN (Global Agriculture Information Network) Report, August 8, http://www.fas.usda.gov/gainfiles/200608/146208611.pdf, last accessed April 1, 2008.

White House. 2007. "President Bush and President Lula of Brazil Discuss Biofuel Technology." Press release, Office of the Press Secretary, March 9. Retrieved December 13, 2007 (http://www.whitehouse.gov/news/releases/2007/03/20070309–4.html).

Zacharias, A. 2005. *Shuck the Sheiks*. New York: Universe.

In: A Sociological Look at Biofuels
M. S. Carolan, pp. 77-79

ISBN: 978-1-60876-708-3
© 2010 Nova Science Publishers, Inc.

EPILOGUE

It's 43 degrees Celsius. A trail of dust rises off the road like steam as the car speeds over the loose gravel. We approach a ridge. Just then it comes into sight. Glistening under the desert sun we see what looks to be water-filled solar panels. In reality it's a massive algae farm. Lying before us are hundreds of rectangles that contain within them thousands of smaller rectangles. Vats of algae as far as the eye can see. We are here to watch algae being processed into biodiesel

This place does not exist (yet?). Research into biofuels derived from algae does, however, represent a hot field for engineers and other scientists at the moment (an area that my own institution, Colorado State University, just happens to be at the forefront of). John Sheehan, the Vice President of Strategy and Sustainable Development at LiveFuels (a firm in Menlo Park, CA specializing in algae biofuel technology), estimates that an acre of algae can produce 50 times more oil than an acre of soybeans (Kanellos 2007). For comparison: soybeans produce approximately 50 gallons of oil per acre per year; canola, 150 gallons; and palm, 650 gallons. In the near future, algae farms are expected to produce 10,000 gallons per acre per year, and in the later future even more (Popular Mechanics 2007). Three ingredients are needed to grow algae: water, sunlight and carbon dioxide. The biomass leftover from the conversion process can be fed to cattle as a protein supplement. And, side-stepping a major criticism of 1st generation biofuels—e.g., corn ethanol and soy biodiesel—algae can be grown on land unsuitable for food production, like a desert.

Of course, there remain enormous technological hurdles that must be overcome, which is why there are no Algae Biodiesel signs at the corner gas station. Algae require significant amounts of water (though after the initial water investment subsequent water requirements will likely be minimal). The temperature of the water has to be just right. Also, the USDA has yet to approve the leftover biomass for cattle feed (animal scientists at Colorado State University are currently in the process of evaluating the safety of algae-based feed on animals). As I write these words, Exxon announced it will invest US$600 million in research devoted to making biodiesel from algae. Money into researching algae-based biodiesel is certainly a necessary step toward making this fuel widely available. However, without accompanying investments in infrastructure, diesel automobiles, and experts systems—in other words, in the broader socio-technical system—all the research in the world will not bring on a substantial shift in the field of transportation.

I make this turn to algae not to promote biofuels derived from this renewable material but to highlight the outside-the-box direction research in this area is taking. For another example: take research occurring at University of Nevada-Reno into developing biodiesel from oil derived from coffee grounds. This research indicates that used grounds are 10 to 15 percent oil by weight. Researchers involved remarked that "there is so much coffee around that several hundred million gallons of biodiesel could potentially be made annually" (though this represents less than 1 percent of the diesel used in the US annually) (as quoted in New York Times 2008). My point: there is no glass ball we can peer into to predict what the future of biofuels will look like. With each passing year we are finding something new that can be processed into fuel.

Just because something is scientifically and technologically feasible from a processing standpoint does not mean its future adoption is probable from a socio-technical point of view. Only a very rudimentary understanding can be achieved when a technological artifact is looked at as an independent "thing", abstracted from its broader system requirements. In the case of coffee biodiesel, an infrastructure will be required to collect grounds efficiently before the fuel could be produced *en masse*. In the case of algae—assuming future large-scale production would occur in uninhabitable land that is also unsuited for food production—its success is predicated upon a massive infrastructural build-up of roads and water pipelines to bring water into and biodiesel out of the worlds' deserts. (Gravel roads to these "farms" like those mentioned at the chapter's beginning would likely be a rarity.) We must also realize that, unlike Europe where diesel automobiles are far more widespread, any major shift to biodiesel in the US would require a shift in consumer preference toward diesel automobiles.

The significance of the socio-technical context for understanding past, present and future biofuel trajectories is a (if not *the*) major take-home point of this book. This also highlights a space that social scientists—like myself—can engage in this discussion. Understanding technological artifacts like fuel as an effect of a complex process—that includes social, political, economic variables—immediately opens debates about its future up to fields other than engineering, petroleum geology, and chemistry.

I recently read an article on *CNN.com/technology* titled "Why our 'amazing' science fiction future fizzled".[1] The article begins by noting how in the 1964 World's Fair people stood in line for hours to see a miniaturized replica of a proto-typical twenty-first century US city, complete with moving sidewalks, jet-packs, and congestion-free highways. The question posed by the author of the article: Why isn't the future what it used to be? If chemurgists from a century ago were given a glimpse of the world today they too would be asking something similar, no doubt expecting that by now we would be living in an agrochemical-based society. The future is not what it used to be because of such phenomena as technological momentum, socio-technical closure, and interpretive flexibility/stability. These sociological facts highlight why radical technological changes do not—and cannot—occur with a mere snap of the figures. Moreover, they reveal the improbability of techno-utopian visions of the future, such as those offered by Jules Verne (author of *Twenty Thousand Leagues Under the Sea*), even though many future technologies of the past have been technologically feasible for quite some time (such as the jet-pack and Smell-O-Vision). Let us not lose sight of these sociological facts as we debate biofuels.

ENDNOTES

[1]

http://www.cnn.com/2009/TECH/science/05/29/jetpack/index.html?iref=newssear

ch

REFERENCES

Kanellos, Michael 2007. The challenge of algae fuel: An expert speaks, *Cnet News* August 23, http://news.cnet.com/8301-10784_3-9765452-7.html, last accessed July 10, 2009.

New York Times 2008. Diesel, Made Simply From Coffee Grounds (Ah, the Exhaust Aroma), *New York Times* December 15, http://www.nytimes.com/2008/12/16/science/16objava.html?_r=1, last accessed July 10, 2009.

Popular Mechanics 2007. Pond-Powered Biofuels: Turning Algae into America's New Energy, *Popular Mechanics* March 29, http://www.popularmechanics.com/science/ earth/4213775.html, last accessed July 10, 2009.

ABOUT THE AUTHOR

Michael S. Carolan is an Associate Professor of Sociology at Colorado State University (CSU). Michael is also a Founding and Core Team Member of the Institute for Livestock and Environment (ILE) at CSU. He received his PhD in Sociology from Iowa State University in 2002. After receiving his doctorate he spent some time at Wageningen University in The Netherlands as a Research Follow in their Environmental Policy Department. From there he held a joint appointment in the Sociology and Environmental Studies Departments at Whitman College in Walla Walla, Washington until moving to Colorado State in 2004. He is currently engaged in three streams of research. One, which is exemplified by this book, examines the socio-technical system of biofuels. His second line of research focuses on patent law, specifically as it applies to biotechnology. Utilizing concepts from the field of Science and Technology Studies, this line of research highlights the work that goes into giving biotechnological artifacts their "objective" qualities (this object-ivity is a requirement for patentability). He is also interested in the relationship between innovation, patents, and development. His third stream of research, grounded in the theoretical framework known as embodied realism, details how changing relationships to food and agriculture have changed our understanding of these phenomena. Along these lines, he looks at how things like community supported agriculture and backyard gardens instill within people certain knowledges that have been lost with the industrialization of the food system. Michael lives just outside of Fort Collins, Colorado with his wife and daughter. His hobbies include hiking, gardening, and visiting with friends and family at local coffee shops.

INDEX

A

AAA, 45
academic, 66
accidents, 92
accounting, 41, 63
accuracy, 42, 96
acquisitions, 94
acute, 22, 73
addiction, 7
additives, 61
adjustment, 51
administration, 60, 68
AEI, 76
afternoon, 1
age, 48
agent, 43, 52
agricultural, 13, 20, 22, 25, 45, 62, 73, 95, 96, 99
agricultural commodities, 13, 20, 22, 73, 96
agriculture, 11, 20, 21, 22, 23, 27, 91, 107
air, 42, 51
alcohol, 3, 4, 9, 10, 11, 12, 13, 14, 15, 16, 20, 22, 23, 24, 25, 26, 27, 28, 34, 39, 40, 42, 43, 44, 46, 47, 49, 50, 51, 52, 53, 59, 72, 73, 81, 96
algae, 103, 104, 105, 106
all-electric, 97

alternative, 6, 16, 52, 59, 65, 66, 69, 70, 72, 79, 93, 94, 96, 97
amendments, 61
american history, 56
analysts, 63
animals, 13, 21, 104
antagonistic, 69
antagonists, 6
appropriations, 99
argument, 1
asia, 81, 87, 97
assessment, 55
atmosphere, 86
attitudes, 69
authority, 61
automobiles, 11, 12, 16, 18, 52, 70, 97, 104, 105
averaging, 91
aversion, 66

B

ballistic missile, 42
bankruptcy, 97
barley, 45
battery, 96
beef, 90
beliefs, 7
benefits, 6, 7, 66, 79, 81, 92, 93

biodiesel, xi, 1, 2, 88, 94, 103, 104, 105
biodiversity, 2, 82, 88
biofuel, xi, 1, 2, 3, 5, 7, 54, 62, 65, 66, 67, 68, 69, 70, 71, 72, 74, 76, 77, 79, 80, 81, 82, 84, 86, 87, 88, 89, 90, 91, 92, 93, 94, 95, 96, 98, 100, 103, 104, 105, 106, 107
biomass, 71, 81, 82, 86, 89, 91, 93, 96, 100, 103, 104
biotechnological, 107
biotechnology, 107
birth, 33
blends, 2, 3, 4, 23, 24, 25, 27, 28, 35, 42, 51, 61, 69, 82, 87, 97, 100
boats, 11, 15, 52
boilers, 9
Brazilian, 74, 89, 91, 94, 99
breakdown, 80
bribes, 13
British Petroleum, 5, 68
burn, 27
burning, 4, 9, 11, 15, 21, 43, 51
bust, 53

C

campaigns, 64
candidates, 64
CAP, 98
carbohydrate, 26
carbon, 4, 43, 76, 78, 86, 98, 99, 100, 103
carbon dioxide, 86, 103
carbon emissions, 86
carrier, 40
cartel, 48, 49, 53
cast, 73
cattle, 103, 104
caucuses, 3
cellulose, 33, 71
cellulosic, 1, 5, 7, 68, 75
cellulosic ethanol, 1, 5, 7, 68, 75
census, 29, 30, 32
CFD, iii
chemical properties, 33
chickens, 27, 70, 71
children, 20

chimera, 42
Civil War, 9
Clean Air Act, 61, 77
closure, 41, 51, 106
CNN, 105
CO2, 83, 86, 87
coal, 11, 15, 33, 54
coal tar, 33
coffee, 2, 104, 105, 108
collaboration, 65
collateral, 66
combustion, 1, 9, 10, 11, 39, 42, 43, 51, 52, 69, 70
commerce, 49
commodity, 15, 21, 22, 26, 45, 46, 73
Commodity Credit Corporation, 45
communities, 89, 107
community support, 107
competition, 13, 15, 41
complement, 6
complexity, 4
compliance, 26
computing, 41
concentrates, xi, 1
conception, 55
confidence, 2, 62
configuration, 47
congress, vi, 10, 23, 63, 69, 78
consciousness, 4
consensus, 86, 87, 89, 91
conservation, 47
consolidation, 14
constitution, 55
construction, 41, 60
consumers, 23, 27, 53, 95
consumption, 5, 10, 16, 73, 96
contamination, 61
contract prices, 26
control, 52
convergence, 42
conversion, 15, 71, 89, 103
corn, 1, 2, 6, 12, 21, 22, 26, 44, 45, 46, 59, 60, 62, 64, 65, 74, 76, 79, 81, 82, 86, 88, 90, 91, 94, 98, 103
Corporate Average Fuel Economy (CAFÉ) 70

costs, 7, 41, 60, 63, 64, 67, 74, 76, 79, 81, 92,
     93, 94, 98
cotton, 10, 45
CRC, 99
credit, 63, 70
criticism, 103
crop insurance, 45
croplands, 100
crops, 21, 22, 28, 45, 98
CRS, 78, 100
crude oil, 16, 18, 47
CTA, 55
cultivation, 91
customers, 24
cutters, 92
cycles, 48, 49, 53

D

debt, 76, 86
decisions, 46
defense, 20
definition, 42
democrats, 64
denial, 72
Department of Agriculture, 5, 65, 68, 78, 94,
     100
Department of Energy (DOE), 5, 68
Department of Justice, 24
deposits, 61
depressed, 73
desert, 103, 104
designers, 71
determinism, 41, 52
diesel, 14, 35, 51, 104, 105
diffusion, 42
direct cost, 63
discipline, 54, 69
discourse, 72, 93
discovery, 56
discriminatory, 13
displacement, 84, 90
distribution, 6, 13, 15, 53, 69
divergence, 80, 88
diversification, 69

diversity, 82, 89, 90, 100
dominant firm, 66
doors, 28
drought, 45
DuPont, 24, 33, 44
dust, 103

E

earth, 106
ecological, 13
economic change, 77
economic efficiency, 41
economic sustainability, 48, 69
economics, 46, 55
economies of scale, 48
ecosystem, 86
egg, 70
embargo, 4, 59
emission, 61, 88, 97, 99
emotions, 7
energy, 1, 2, 5, 20, 40, 61, 63, 70, 72, 74, 75,
     84, 89, 91, 96, 97, 98, 99
Energy Information Administration, 63
Energy Policy Act of 2005, 61
engines, 3, 11, 39, 40, 42, 44, 51, 52, 53, 69,
     71
environment, 22, 66, 88
environmental impact, 6
environmental policy, 98
Environmental Protection Agency (EPA), 61,
     70
environmental sustainability, 96
enzymes, 71
erosion, 84, 91
ether, 61
ethyl alcohol, 1, 2, 3, 4, 9, 12, 26, 33, 39, 48,
     59, 60, 71
Europe, 10, 19, 21, 105
European Union, 96
evil, 12
excise tax, 63
expansions, 94
expert systems, 6, 65, 66, 67, 80
expertise, 65, 66

extraction, 46, 48
eyes, 9

F

failure, 43
family, 108
farmers, 21, 22, 26, 39, 45, 46, 63, 73, 86
farming, 46, 77
farms, 26, 29, 30, 31, 46, 103, 105
fear, 42
federal government, 10, 23, 63, 97
feedback, 66
feet, 4, 40, 48, 67
fertilizers, 22, 88
financial support, 26
firms, 15, 46, 48, 65, 66
first generation, 93, 95, 96
first world, 21, 26
flex, 70, 77, 92
flexibility, 41, 42, 53, 71, 106
flood, 21
flow, 74, 96
fluctuations, 49
fluid, 25
focusing, 92
food, 2, 7, 13, 35, 74, 77, 81, 90, 98, 104, 105, 107
food production, 104, 105
fordism, 56
fossil, 5, 62, 89
foul smelling, 51
framing, 40, 72
freedom, 4, 43
fuel oil, 14, 15
fuel type, 15
funding, 6
funds, 26
furnaces, 15

G

gas, 2, 12, 13, 26, 28, 33, 39, 40, 54, 59, 69, 71, 74, 104

gasification, 5, 68
gasoline, 1, 2, 3, 4, 12, 13, 15, 21, 23, 24, 25, 27, 28, 39, 40, 42, 43, 44, 47, 49, 50, 51, 52, 53, 59, 60, 61, 63, 65, 69, 70, 71, 72, 77, 81, 82, 86, 88, 91, 100
gauge, 40
General Accounting Office, 5, 8, 62, 76
General Motors, 4, 24
generation, 7, 94, 103
geology, 48, 66, 105
glass, 25, 104
global climate change, 1, 5, 70, 72
global trade, 96
global warming, 73, 86
goals, 72
gold, 7
government, 5, 10, 11, 16, 21, 23, 24, 26, 28, 40, 46, 59, 62, 63, 67, 70, 95, 97
grain, 12, 13, 21, 22, 23, 26, 28, 36, 44, 45, 73, 86
grains, 90
grants, 5, 68
grassland, 82, 86, 89, 100
Great Depression, 22
Great War, 19, 21
greenhouse, 73, 86, 97, 99, 100
greenhouse gas, 73, 86, 98, 99, 100
ground water, 61
grounding, 7
groups, 23, 42, 64
growth, 69

H

harvest, 21, 92
heart, 3
height, 4
herbicide, 71, 88
highways, 105
horses, 21, 46
host, 41, 73
house, 57, 99
household, 59
human, 13, 21, 46, 89, 92
human rights, 92

hydrocarbons, 88

I

ideology, 60
imagination, 27, 33, 53, 54
importer, 19
imports, 16, 65
incentives, 21, 22, 23, 45, 46, 68, 90
income, 21, 53
increasing returns, 6, 55, 67
incubation, 4, 60, 65, 70, 97
independence, 1, 5, 72, 74, 96
indication, 46
Indigenous, 92, 100
industrial, 3, 10, 11, 12, 13, 15, 20, 48, 94
industrial application, 20
industrial production, 94
industrialization, 107
industry, 2, 3, 5, 7, 11, 12, 13, 14, 15, 20, 21,
    22, 23, 24, 41, 44, 47, 48, 49, 51, 53, 60,
    62, 63, 65, 69, 71, 72, 73, 96
inertia, 66
infancy, 15
infrastructure, 47, 53, 104, 105
inhibitors, 7, 75
injury, vi
innovation, 43, 75, 107
instability, 1, 25, 73
institutions, 65, 66
insurance, 39, 45
interaction, 76
interface, 41
internal combustion, 1, 11, 39, 42, 43, 52, 69,
    70
Internal Revenue Service, 9
interstate, 48, 49
inventories, 66
investment, 14, 46, 94, 104
investors, 62, 90
isolation, 40, 47

J

judgment, 79
just deserts, 95
justification, 25

K

kerosene, 11, 13, 14

L

labor, 46
land, 21, 46, 82, 86, 87, 89, 90, 91, 100, 104,
    105
land-use, 82, 86, 87, 89, 100
language, 53
large-scale, 19, 105
law, 10, 12, 22, 23, 107
learning, 4, 67, 94
legislation, 10, 23, 24, 60
lenders, 63
lens, 22, 93
licenses, 24
life cycle, 86, 87, 88, 91, 98
links, 80
listening, 64
livestock, 107
loans, 45, 60, 66
lobbying, 23, 48, 63, 64, 65
locomotion, 11
losses, 90, 95

M

machinery, 11, 16, 66
magazines, 11
magnetic, vi
mainstream, 60
maize, 71
major cities, 13
management, 76
manufacturer, 51

manufacturing, 10, 44
market, 4, 13, 14, 15, 21, 22, 23, 24, 25, 26,
    28, 33, 34, 36, 40, 44, 47, 48, 49, 53, 60,
    62, 63, 65, 66, 71, 73, 95, 96
market opening, 53
market prices, 22, 26
marketplace, 40
marriage, 20
martial law, 23
mask, 92
meanings, 42
measures, 41
men, 21, 22
mergers, 13
metallurgy, 20, 33
metaphor, 52
methanol, 61
microbial, 71
Middle East, 73
military, 11
minimum price, 21
missions, 86, 87
mixing, 12, 61
models, 52
molasses, 82, 90
momentum, 3, 6, 40, 41, 52, 63, 65, 73, 93,
    95, 96, 105
money, 64, 68, 97
monopoly, 44
morning, 2
movement, 4, 14, 21, 22, 23, 26, 27, 48, 61
MTBE, 61, 78

N

nanotechnology, 5, 68
nation, 23, 45, 70
National Research Council, 20
natural gas, 33, 54
navy, 15
network, 5, 13, 42, 55, 65, 69, 95
non-human, 89
non-renewable, 72
normal, 45
nuclear power, 5, 62

O

obsolete, 66
octane, 51
odors, 39
OECD, 96
oil, 1, 3, 4, 7, 9, 10, 11, 13, 14, 15, 16, 18, 19,
    23, 24, 40, 46, 47, 48, 49, 50, 53, 59, 69,
    70, 72, 73, 81, 88, 90, 94, 99, 103, 104
oil production, 14, 49, 59
online, 13
OPEC, 4, 59
optimism, 33, 54
organic, 54
overproduction, 21, 73

P

palm oil, 81, 88
parents, 2
partnerships, 69
patents, 26, 44, 107
peer, 1, 67, 79, 104
penalties, 70
per capita, 59
perceptions, 54
pesticides, 88
petrochemical, 89
petroleum, 8, 11, 14, 15, 16, 19, 23, 25, 33,
    34, 37, 42, 47, 54, 55, 57, 59, 66, 69, 72,
    76, 77, 105
petroleum products, 14
philosophy, 41
pipelines, 15, 61, 105
pitch, 33, 54
planning, 4
plants, 60, 64, 71, 92, 94
plastics, 33, 54
play, 52
politicians, 20, 25, 72
politics, 55
pollution, 61, 73, 89, 92
poor, 77, 86, 100
population, 46

potatoes, 10, 12, 26

power, 2, 5, 10, 16, 21, 27, 39, 42, 46, 51, 53, 62

preference, 40, 105

premium, 27

presidency, 64

president, 12

pressure, 21, 90

price effect, 48

prices, 2, 5, 21, 22, 23, 26, 35, 45, 47, 48, 49, 59, 72, 73, 74, 86, 90

pricing policies, 13

producers, 6, 64, 90, 95

production, xi, 1, 2, 5, 6, 10, 12, 13, 18, 21, 26, 33, 45, 47, 48, 51, 53, 61, 62, 63, 64, 65, 67, 68, 70, 71, 73, 76, 82, 89, 90, 91, 92, 93, 94, 96, 98, 99, 105

production costs, 67, 74, 76, 94, 98

profit, 44, 64

profitability, 15

program, 23, 45

programming, 41

property, vi, 46

property owner, 46

protection, 61, 95

protein, 26, 103

public, 4, 6, 11, 16, 20, 54, 61, 66, 72, 73, 74, 92, 93

public health, 61

public opinion, 74

public support, 11, 73

pumps, 12

pyrolysis, 5, 68

R

rail, 13, 41

rainfall, 91

range, 76

ratings, 70

raw materials, 21, 33, 40, 54

realism, 107

reality, 1, 6, 40, 46, 71, 92, 93, 103

recession, 74

refineries, 15

refiners, 14

refining, 14, 48

regular, 60

regulations, 26, 46, 61, 74

relationship, 107

renewable resource, 9

republicans, 64

research and development, 15, 78

reserves, 1, 11, 16, 19, 43, 45, 59

reservoir, 47, 48

residuals, 14

resilience, 71

resistance, 23, 95

resolution, 51

resources, 9, 15, 62

returns, 6, 14, 80

revenue, 9, 26

RFA, 63

rhetoric, 64, 71

risks, 11, 92

rolling, 13

routines, 66, 71

rubber, 33

rural, 27

S

safety, 73, 104

sales, 16, 28, 44, 60

salt, 9

satisfaction, 21

saturation, 16

scale economies, 5

school, 65

scores, 27

search, 79, 80

second generation, 93, 95, 96

Secretary of Agriculture, 3, 22, 27

security, 89, 90, 98

seed, 22

selecting, 27

self, 6

senate, 11, 64

separation, 25

services, vi

shape, 41, 48, 63, 72
shaping, 52
sharing, 96
Sherman Act, 35, 55
shipping, 13, 48
shock, 51
signals, 61, 62
signs, 104
skills, 66
smog, 61
smoke, 9
smoothness, 27
social construct, 41
social costs, 92
social group, 42
social security, 92
social theory, 6
sociological, 1, 7, 40, 106
sociologists, 40
sociology, 75, 79
soft money, 64
soil, 86, 91, 92
soil erosion, 91, 92
solar, 103
solar panels, 103
South Asia, 81, 87
soy, 104
soybeans, 2, 44, 45, 82, 88, 103
species, 82, 88
species richness, 82
speculation, 7
stability, 41, 42, 45, 48, 51, 53, 65, 67, 106
standards, 23, 70
starch, 45, 71
statistics, 19
stock, 7
storage, 15, 25, 61
stoves, 11
streams, 66, 107
structuring, 52
subsidies, 5, 6, 7, 59, 60, 62, 64, 95
sugar industry, 65
sugarcane, 1, 26, 65, 71, 74, 81, 82, 89, 91, 92, 94
suicide, 3

summer, 74
sun, 35
sunlight, 103
superiority, 3, 50
supply, 12, 13, 16, 19, 72, 90, 95
suppression, 13
supreme court, 24
surplus, 20, 25, 28, 59
sustainability, 48, 69, 74, 76, 96
SUV, 97
switching, 93
symbolic, 60, 72, 73, 74
systems, 41, 66, 67, 69, 70, 72, 104

T

tanks, 25, 61
targets, 92
tariff, 65, 74
tax credit, 60, 63
tax incentives, 23
taxation, 10, 47
taxes, 11, 40
taxpayers, 64, 95
technocratic, 33, 53
technological change, 4, 40, 62, 106
technological developments, 78
technology, 14, 21, 26, 40, 41, 42, 52, 55, 57, 62, 76, 79, 95, 96, 103, 105
temperature, 104
tension, 52
territory, 24
thermoplastic, 33
thinking, 65
threatened, 24, 61, 70, 77, 90, 95
time, xi, 1, 2, 3, 6, 9, 10, 11, 12, 14, 15, 16, 19, 20, 21, 26, 40, 42, 43, 44, 45, 46, 47, 52, 54, 66, 71, 74, 94, 95, 106, 107
timing, 100
title, 3, 15, 80
total product, 10
traction, 72, 73
trade, 96
trade agreement, 96
traffic, 66

trajectory, 9, 41, 52, 66, 96
transaction costs, 41, 60
transformations, 52
transition, 7, 60
transport, 24
transportation, 41, 53, 66, 73, 74, 97, 104
trees, 5, 68
trucking, 41
trucks, 2

U

U.S. Department of Agriculture (USDA), 5, 7, 26, 28, 37, 45, 65, 68, 75, 78, 94, 104
unfolded, 1
United Nations, 92, 100
universities, 67

V

values, 98
variability, 41
variables, 92, 93, 105
vegetable oil, 1
vehicles, 70, 97
venture capital, 67
vice president, 69, 103

W

war, 11, 16, 19, 21
War of 1812, 10
waste products, 86
water, 4, 9, 16, 24, 25, 61, 89, 103, 104, 105
watersheds, 89
waterways, 89
welfare, 7
wells, 13, 48
wheat, 21, 36, 45, 82
white house, 3, 96, 101
wholesale, 24
wind, 91
winter, 1, 91
women, 22
wood, 12
workers, 92
World War, 4, 11, 16, 20, 21, 25, 26, 33, 35, 47, 54, 59, 73
World War I, 4, 11, 16, 20, 21, 25, 33, 35, 47, 54, 59, 73
World War II, 4, 33, 47, 54, 59
writing, 43

Y

yield, 14, 16